U0215506

|引领时尚潮流|

深圳视界文化传播有限公司 编

设计思维

服装店面设计与陈设

FASHION STORE DESIGN AND FURNISHING

中国林业出版社
China Forestry Publishing House

PREFACE

序言

探寻内心深处的设计思维

EXPLORING THE DESIGN THOUGHTS IN THE DEEP HEART

从每个项目成形的地方开始，我们的设计都始于一个坚实概念的设想。我们的目标是寻找一些能够激发创意的方式，而这些创意是足够扎实的，它们能够支撑我们完成整个设计项目。这些创意和概念可以来自于在不同地方获得的不同启发，它们甚至可能与项目的主要性质无关，可以像艺术或生活一样丰富多彩。

值得注意的一点：当前的社会环境中，人们强烈要求将体验去机械化，希望所有事物都能够以更人性化、更温暖、更富于情感的形式成为一种服务，成为一种他们能够购买的体验式服务。

一个经过设计思考的空间会自动地从根本上转化为一种体验的本身，或者至少成为这种体验的绝大部分。设计思考的主要动机之一，是有意地引导使用者去感知和融入空间，达到“将经验去机械化”的目的。许多人仍然没有理解的是，空间的规划设计在这方面可以是一个关键的变量，它能够使整个空间真正发挥其实际作用。

大多数的情感经营需要付出真正的努力，而不仅仅是简单的金钱交易关系。实际上你可以认为这是一个时机，你可以借机环游世界，去看看已经定型的或即将到来的趋势，去接触一些不一定适合你的商店、空间，或者一些你不熟悉的人，从内心出发、从全人类出发、甚至是从脆弱的自己出发，引导你在任何事情上都做出一个最真诚的表达，做出发自内心的设计，成就人性化、温暖、赋予情感的设计空间。

需要注意的另一方面是：时下我们存在于一个非常特殊的环境中，这个环境具有自动更新的本质，而这种“更新”发生的频率极高，它会使许多曾经有用的或是必需的东西变得过时，甚至变成你的设计作品中的败笔。就零售商店的设计而言，有些设计方案乍一看去是值得怀疑的，但仔细思考后会发现这些可疑的方案却成了唯一可行的，有时甚至是最佳的可行方案。

由此看来，现在的我们比以往任何时候都更需要注意到不断变化的环境，尤其需要注意用户的体验感。作为设计师，我们的主要目标应该是在自身中生成一个不断自动更新的系统。你不需要成为这方面的专家，需要的只是一双处处留心的眼睛。

尽管目前似乎没有我们可以依赖的具体手册或指南，而且我们需要通过内省来为设计做出更多努力，但这是一个很好的机会，可以将更多自己的感悟和特色注入作品中。

零售商店的设计也不例外。我们发自内心地感知世界，将感知到的东西转化为独特的创意，再像一位手艺人一样，雕琢出具有品牌特色的销售

Every project we design starts with the conception of a solid concept from where everything starts shaping. The objective is to find ways to inspire creations that are really solid to give us supports to develop and finish the whole project. These creations and concepts can come from a wide variety of places and inspirations. They might even not be associated with the main nature of the project in turn. It can be as varied as art or everyday life.

Notably, the current social climate has people screaming for the de-mechanization of experience in favor of a more humanly warm and empathic sympathy in absolutely everything, becoming an experiential service that they could be paying for.

A design-thought space will automatically and radically transform into the experience itself or at least great part of it. One of the main motors of design thought is to guide its occupants to sense and fuse space, achieving the aims of "de-mechanization". What many people still fail to grasp, is that space design can be crucial variable that can make the whole planed use for the same space to actually work.

The sentimental approach of today involves a real effort and not just a simple appropriation. You can actually recognize the unique opportunity that you are presented with. Actually, you can take a look around the world, see the upcoming and already positioned trends and get in touch with some store, space, or some of your unfamiliar people that may not necessarily suit you, and instead, choose from your heart, from human beings and even the vulnerable part of yourself. This will only result in a completely honest expression on anything you design, creating a humanized, warm and emotional design of space.

In the other hand, nowadays, we are immersed in a really special situation, which presents itself with an auto-regenerative essence, occurring in a constant frequency. It makes everything that was useful or even a necessity before become something out of style or something that can determine the downfall of whatever it is that you are designing. Speaking solely about the retail store design, there are various examples out there of questionable solutions at the first glance, but that when thought carefully, they suddenly appear as the only viable one and even the best move possible.

Therefore, we need to be aware of the ever-changing environment with a special attention to the user experience more than ever. As designers, our main goal should be to generate a constant auto-update flow in ourselves. You do not need to be an expert on the subject, all you need is a pair of watchful eyes.

Even though there is no current specific manual or guide that we can rely on and we need to make an extra effort for design through introspection, this is a great opportunity to actually inject a little more of your understands and features into your work.

And the retail store design is no exception. We perceive the world from our heart, transform what we perceive into the unique ideas and design a sales space with the brand characteristics like a craftsman.

Anagrama

CONTENTS

目录

CHAPTER 1
The Way to Do the Storefront Design Well
如何做好服装店面设计

CHAPTER 2
Detailed Analysis of Global Projects
全球实景项目详解

016 The "Fish-eye" Construction to Show the Pure Self of the Dark Style
鱼眼式构造，透视暗黑风格的纯粹自我

030 The Ornate Area of Luxury Clothing
奢侈品服饰的华丽据点

048 Reproduce History and Reappear Modernity
重塑历史，再现摩登

064 Nike Shanghai Marathon Expo 2017
2017 上海国际马拉松 · 耐克

080 The Street Store, The Contrasting Fashion
街边的店铺，对比的时尚

090 Plainness but Exquisiteness and Elegance
质似朴，实则细腻又优雅

098 Abandon Inefficiency, Explore the Interest of Structure
抛除低效，探寻结构趣味

110 *The Perception of Future*
感知未来

122 *A Figured Lady·A New Fashionable Style*
百变女郎・穿出新时尚

140 *Displaying Youth and Personality*
张扬青春个性

150 *The Women's Fashion Shop With A Mobile Amusement Park*
装着流动游乐园的女装店

162 *The Elegant Pink Dancing With the Unruly Geometric Elements*
高雅粉与几何元素的不羁共舞

170 *The Planet of A Tender Idealist*
一个温柔理想主义者的星球

180 *Wandering in the Bright Shadow*
游走于明净与光影照拂之中

188 *The Children's Clothing Store Encountering a Shark*
撞见鲨鱼的童装店

202 *The Dreamy Architecture at the Corner, the Charming Rock Style*
街角的梦幻建筑，迷人的摇滚风

214 *Extreme High-end Black, the Unusual Artistic Style of A Small Store*
极致高级黑，小店不凡的艺术格调

224 *CARMEN: The Shop of Time*
卡蔓：时光店铺

234 *The Modern High-end Gray, The New Interpretation of Freedom and Humanity*
现代摩登的高级灰，自在与人文相融的新演绎

244 *A Extravagant Montage*
一场奢华的蒙太奇

254 *Attribution and Stimulation Under the Guidance of Artistic Magic*
艺术魔力指引之下的归属与刺激

266 *The Fashionable "Upside Down"*
时尚"逆世界"

276 *The Modular "Lilong"*
购物空间的模块"里弄"

286 *The International Fashion Rooted in the Chinese Culture*
根植于中国文化的国际时尚

294 *The Office Lady · The Extraordinary Fashion*
职场女性・非凡时尚

300 *Enjoying the Leisure of Suzhou Garden in AKIPELAGO*
集岛成群，享苏州园林式悠游

310 *A Three-dimensional Magazine, A Living Store*
立体的杂志，活着的店铺

1
CHAPTER

THE WAY TO DO THE STOREFRONT DESIGN WELL

如何做好服装店面设计

THE WAY TO DO THE STOREFRONT DESIGN WELL

如何做好服装店面设计

论 | 述 | 篇　Statement

基 | 础 | 篇　Fundamental

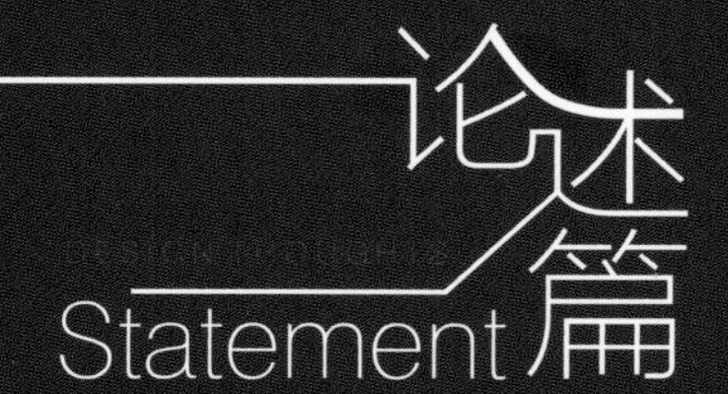

随着互联网经济时代的快速发展，实体店面设计的升级迫在眉睫。好的店面设计，可以展现品牌的独特魅力，是顾客在瞬间判断一家店铺形象所凭借的依据。良好的第一直观印象是顾客驻足的关键，吸引顾客入店，则是促进成交不可或缺的因素。

纵观当今的服装市场，服装品牌之多，除了在服装自身设计和品质上的竞争以外，服装销售终端的竞争也是愈演愈烈。如果服装的店面设计不能给经过的人视觉美感，就等于没有了自己的卖点。因此，在同质化如此严重的当下，服装品牌的推广显得极为重要。各大商家都在极力创新服装店面设计，希望可以设计一家独具特色的店铺，通过店面升级给品牌带来更大的效益。而服饰产品是典型的具有强烈“触摸”和“试穿试戴”体验的品类，如何让消费者能够更乐于走近服装、欣赏服装、认同服装，从而促成购买服装这一行为，是服装店面升级最终的目标。

优质的服装店面设计能在第一时间影响到置身于这个环境中的人，它不再仅仅是一家销售性质的店铺，而是一个集体验、情感与文化为一体的综合媒介。它既能满足消费者热爱时尚与讲求个性的物质需求，又能及时提供多元化的潮流讯息，还可以满足消费者精神上的愉悦需求。

因此，让人眼前一亮的服装店设计是服装销售必不可少的前提。

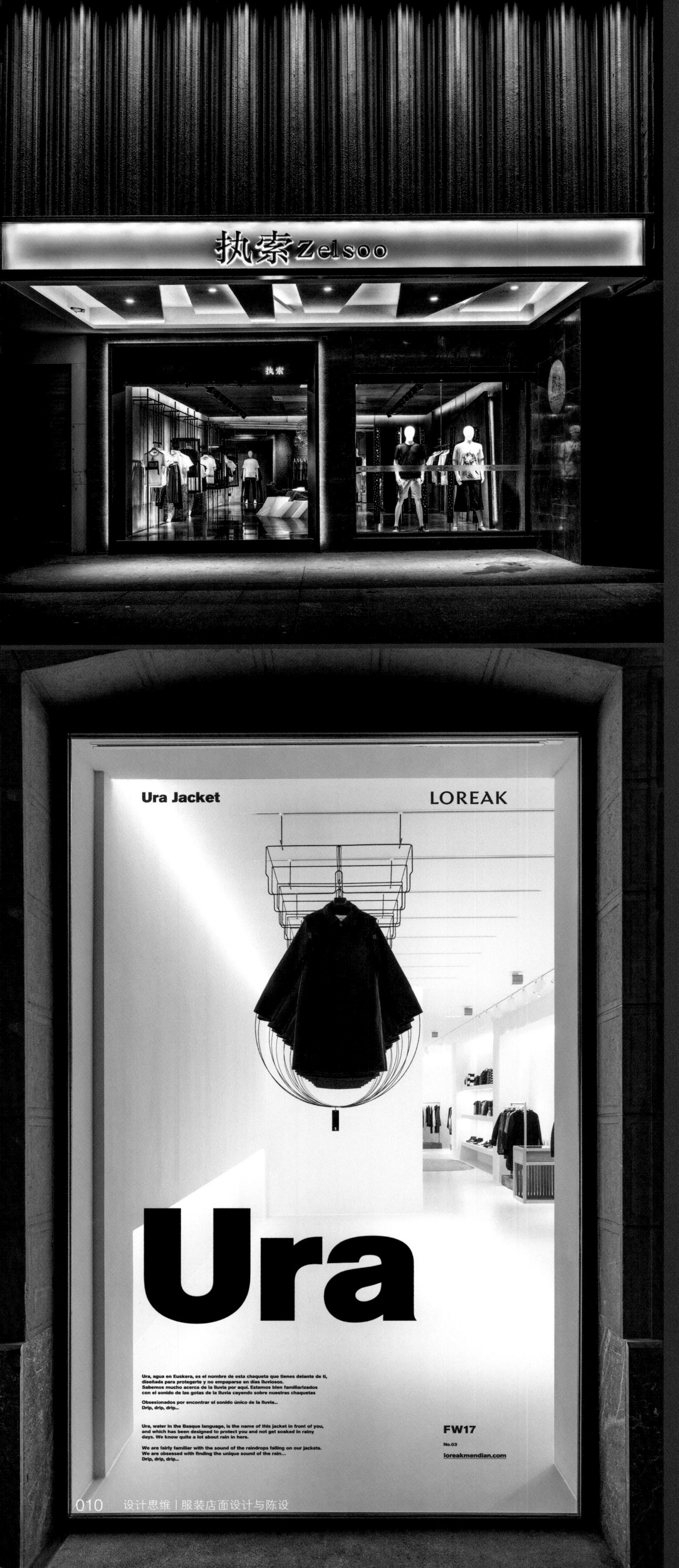

基础篇 Fundamen

DESIGN THOUGHTS

很多人都清楚服装店面设计对于服装店的重要性，但是具体如何做好一家服装店面的设计？如何精确地用设计去诠释这家服装店的品牌、风格以及形象？设计师可以从以下这些方面思考，轻松掌握店面设计的精髓，并在实践中精准打造出适合各大品牌所追求的店面形象。

外观招牌设计 01

Exterior Signboard Design

外观招牌设计是店铺给消费者传递的第一印象，作用是吸引视线，使人们对门店产生兴趣，并进一步激发他们想进店看一看的参与意识，同时这也是自身品牌形象的基本确立，体现服装店铺的品位和个性。在设计时，应力求多样化和与众不同，既要做到引人瞩目，又要与店内设计融为一体，给人以完美的外观形象。

外观类型分为全封闭型、半开放型、敞开型。而设计师如何在众多类型之间做到设计的完美，具体可从以下几个方面入手：

1. 视觉上新颖美观、醒目大方，让顾客或过往行人都能以较远或多个角度清晰地看见。

2. 在构图、造型、色彩、比例等设计方面，力求言简意赅、易读易记、富有美感，具有较强的吸引力，给消费者以良好的审美感受。

3. 在材质上的精准选择，选用适合营造品牌形象气质的材料，如木质、石材、金属材料等，制作成与经营内容相一致的形象或图形，增强品牌的直接感召力。

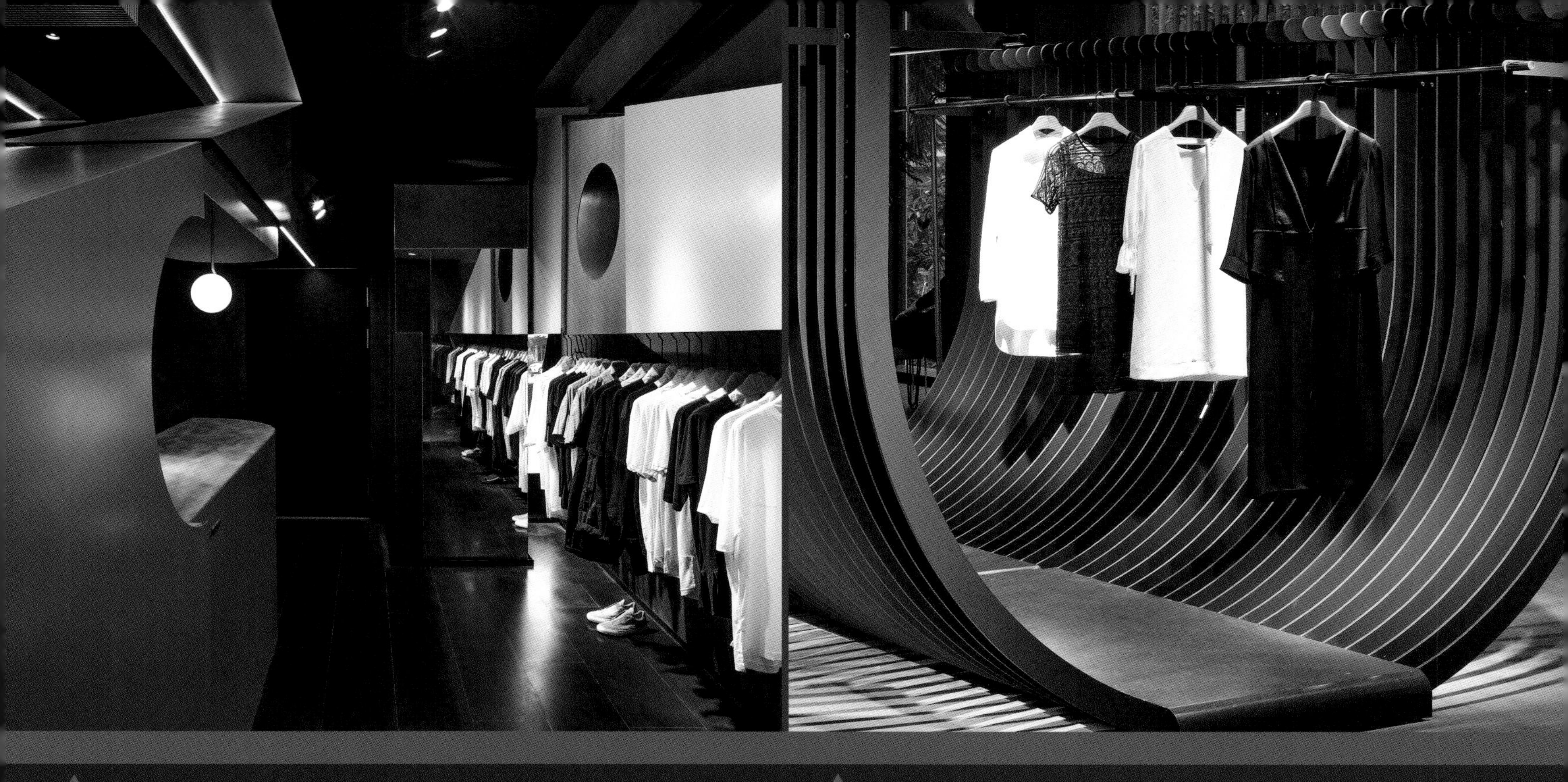

▲
直线式布局

▲
陈列式布局

02 空间布局设计

Spatial Layout Design

直线式布局 | 岛屿式布局 | 斜角式布局 | 陈列式布局 | 格子式布局

一家服装店的布局所影响的效益短时间内不会体现出来，但时间一长，便能明显看出服装店布局对店面相关效应的强烈影响。所以，服装店布局的重要性不可忽略，只有有效合理的服装店布局才能恰当地展现出服装产品的特性、质感和理念，让消费者看到产品全面的信息，了解产品的功能，刺激消费者的购买欲望。

从消费者的右行习惯、便利性、舒适性等方面考虑，服装店铺常用的 5 种布局类型有：

1. 直线式布局。这种布局又称沿墙式布局。在这种布局中，柜台、货架都沿墙成直线摆设。这种形式不受营业场所大小或墙角弯度的限制，能够展示较多的商品，是最基本的设计形式。直线式布局的优点在于能够方便店员拿取商品，能够随时补货，有利于节省人力。

2. 岛屿式布局。这种布局是指柜台以岛状分布，四周用柜台围成封闭状，中间设置货架。在布局时，可以摆设成圆形、长方形、三角形等。这种布局能够充分利用室内光线和空间，为卖场争取到更多的有效面积。基于岛屿自身的形状，它能随地形和营业场所支柱等情况来装饰店铺空间，起到美化的作用。但缺点在于，不利于上货补货，且面积有限，所能陈列的商品不多。

3. 斜角式布局。该布局是指利用店内的设备和建筑空间，如柜台、货架等与室内的柱子围成斜角形状的布置。它能为室内增加延伸的视觉效果，让内部布局变化具有空间性。

4. 陈列式布局。这种类型一般是指在营业场所中央，设置若干陈列柜、货架等，展示各种商品，前边摆设若干柜台进行销售。在这种布局里，店员的工作区域和顾客区域重合，两者都在同一区域活动，可以活跃卖场的人气，形成互动的卖场氛围，也有利于提高服务质量，是一种比较自由、灵活的设计形式，逐渐被广大服装店经营者采用。

5. 格子式布局。该布局类型结构严谨，是一种十分规范的布局方式，能够轻易博得顾客的信任。一些服装店利用格子式布局，将所有的柜台摆放成互为直角的关系，构成曲径式通道，形成客流动线。这种布局，能产生一种动力效果——人流由入口经过商品展示区，然后通向店铺出口，给人以井然有序的印象。同时，格子式布局大多用于敞开式销售，能让顾客自由选择，满足现代顾客对自由、闲适的购物环境的追求。

主题陈列、定位陈列

整齐陈列

整体陈列

室内陈列设计 03

Indoor Furnishings Design

整体陈列 | 主题陈列 | 整齐陈列
定位陈列 | 关联陈列 | 分类陈列

服装店陈列是一门学问，那么，如何陈列能够产生美感，怎样的陈列才能吸引顾客并让顾客赏心悦目？以下是服装店陈列的 6 大原则：

1. 整体陈列。将整套商品完整地向顾客展示，如将全套服饰作为一个整体，用人体模特从头至脚完整地进行陈列。整体陈列的方式能为顾客作整体设想，便于顾客的购买。

2. 主题陈列。创造独特的氛围，吸引顾客的注意力，进而起到销售商品的作用。

3. 整齐陈列。按照货架的尺寸、商品的数值，整齐地排列，突出商品的量感，从而给顾客一种购买刺激。

4. 定位陈列。指某些商品一经确认陈列位置后，通常不再作变动。需定位陈列的商品一般是知名度高的名牌商品，顾客购买这些商品频率高、购买量大，因此需要对这些商品给予固定的位置来陈列，以方便顾客，尤其是老顾客。

5. 关联陈列。关联陈列指将不同种类但相互补充的服饰陈列在一起。运用商品之间的互补性，能够使顾客在购买某商品后，也顺便购买旁边的商品。它能够使得服装店的整体陈列多样化，也增加顾客购买商品的概率。

6. 分类陈列。根据商品质量、性能、特点、颜色和使用对象等进行分类，方便顾客挑选和比较。

灯光照明设计 04

Lighting Design

灯光照明设计分为 3 个层面：第一层是基本照明，即保证店铺起码的能见度，方便顾客选购商品；第二层是商品照明，为了突出商品的特质，吸引顾客的注意而设置的特殊照明；第三层是装饰照明，可以营造店铺独有的形象与气质，且装饰灯光的设计与商店整体形象是协调一致的。因此，灯光除了基本的照明功能外，还能表现商品特质，突出品牌特性，呈现陈列的最佳视觉效果，从而促进互动体验，实现销售。

下面具体分析各区域灯光照明设计的目的和原则：

外部灯光：它负责照亮店门和店前环境，起到渲染店铺气氛、烘托环境、增加店铺门面形式美的作用。此处大多使用人工光源与色彩进行搭配。色彩是人视觉的基本特征之一，不同波长的可见光引起视觉对不同颜色的感觉，形成不同的心理感受，从而吸引顾客进店。

橱窗灯光：为了达到展示品牌风格、季节服装特色、服饰主题等效果，此处灯光安排原则以创造吸引力、重点照明、灵活变化为主，大多采用垂直照明。

销售区域灯光：这里是服装店灯光设计的重点区域之一，照度要高于其他照明区域，除了要满足基本照明的功能之外，还要注意营造情感和气氛，强调品牌信息。此处大多会重叠使用灯光，有点有面，宽窄结合，层次丰富。

陈列区域灯光：这个区域也是服装店铺灯光设计的重点区域，是商品卖点的主要展示区域，是店铺实现销售的主要视觉冲击区。灯光的主要任务是营造好的灯光环境，形成有效的视觉冲击，能有效地表现服装特质，吸引顾客体验、互动和购买。照度同样会高于其他照明区域，重叠使用灯光也是这一区域的表现手法，通常是一个流水台配备一盏灯照射。通过灯光的投射，可以使服装产生不同的明暗效果，更有立体感、材质感，并获得展陈所需的照明气氛，突出品牌特质。但是光线由于光束角和投射方向的不同，产生的效果是截然不同的，正确使用灯具和灯光组合是达到展陈和照明目的的关键。

陈列区域灯光

CHAPTER 2

DETAILED ANALYSIS OF GLOBAL PROJECTS

全球实景项目详解

THE "FISH-EYE" CONSTRUCTION TO SHOW THE PURE SELF OF THE DARK STYLE

鱼眼式构造，透视暗黑风格的纯粹自我

品牌文化 Brand Culture

|GYWJ|

Founded in 2013, GYWJ is a rare designer clothing with dark style, taking the deconstructionism as aesthetics, the dark black as the based tone. GYWJ believes that the highest level of clothing and dressing is the personality and elegance, fashion and fascination that are shaped under the normal conditions. Its clothing emphasizes the naturalness and simplicity and pursues a harmonious state between the human body, clothing and environment. The casual style of clothing, the imported apparel fabric and the perfect craftsmanship create comfort in wearing, express the unfettered attitudes and woman's stretching postures, convey the core culture of "comfort, stretch" and the life aesthetics of "Give Yourself Wonderful Journey".

项目信息
PROJECT INFORMATION

扫码查看电子书

项目地点—浙江杭州
设计师—叶梹
设计公司—叶梹室内设计工作室
项目名称—GYWJ（己以）

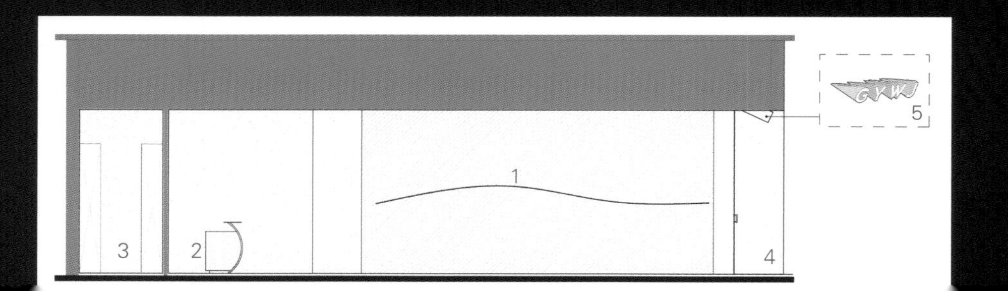

SECTION B

1 Sales Area 销售区
2 Checkout Counter 收银台
3 Facade 橱窗
4 Entrance 入口
5 Modeling Analysis 建模分析

GYWJ创立于2013年，是国内少有的以解构主义为审美，以暗黑色为主基调的暗黑风格设计师女装品牌。GYWJ认为穿衣打扮的最高境界是常态下塑造的个性与优雅、时髦和迷人，其服装讲究自然、简约，追求人体、服装和环境之间的和谐状态。随性的服饰风格、进口衣料和完善的工艺，创造出穿着的舒适性，表达出无拘无束的态度和女性舒展的姿态，传递“舒适、舒展”的核心文化和“Give Yourself Wonderful Journey”的生活美学。

In the commercial space of Hangzhou landmarks with numerous brands, this project has an ideal space layer. Therefore, the designer hopes to adopt the "fish-eye" structure and gets through the connection of space to raise the attention of the entire space. Corresponding to the deconstruction of the GYWJ brand aesthetic, the whole space is concentrated, hierarchical and harmonious, uses the dark black as the main color tone, and the custom-made iron, custom paint technology to show the high-end, personalized decoration style. The use of color and materials is extremely simple and full of postmodernism. Looking at space, it seems to have an unspeakable sense of future. When you enter into space, you will be intoxicated in the low-key atmosphere and return to the self-follow.

SECTION A

1. Sales Area 销售区
2. Checkout Counter 收银台
3. Plastic Coated Steel Wire 涂塑钢丝

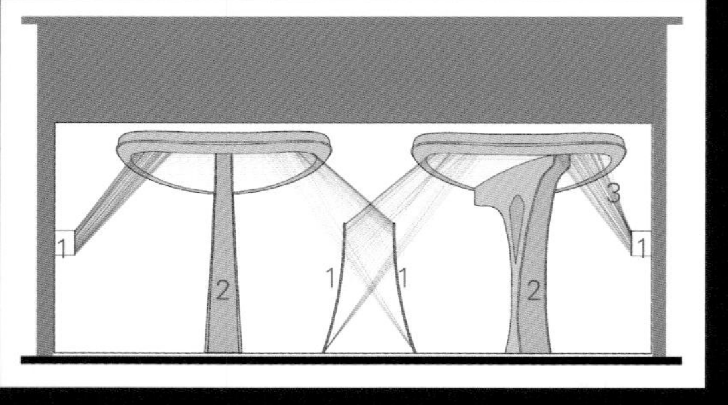

1 | 2
3 4

在品牌林立的杭州地标性商业空间来福士中，本案拥有较为理想的空间层高，由此，设计师希望采用“鱼眼式”结构，同时打通空间连接，提升了整体空间的关注度。对应GYWJ品牌的解构主义审美，整个空间集中、有层次、彼此和谐，以暗黑为主色调，以定制铁艺、定制油漆工艺等展现出高端、个性的装修风格，用色与使用材料都极为简洁，充满后现代主义风格的气质。放眼空间，似有一种不可言说的未来感，进入空间，则会沉醉于低调，回归到对自我的追随中。

Overall Structural Analysis

1 Major Sales Area 主销售区
2 Entrance 入口
3 Facade 橱窗

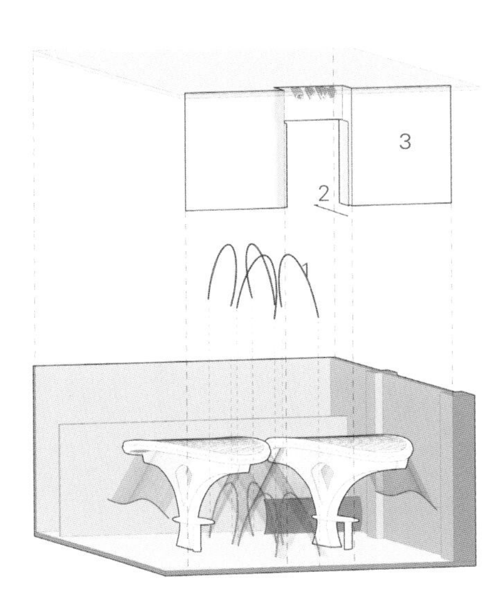

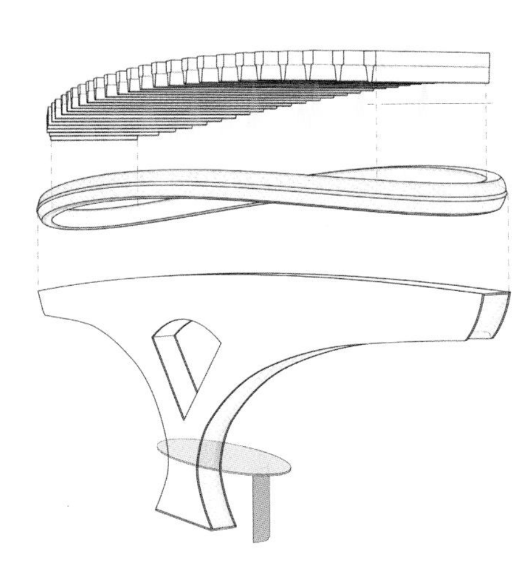

室内陈设
Indoor Furnishings

[1] 强化中岛部分，提升空间连贯度

设计师运用鱼眼式的构造强化出中岛部分，用钢丝营造面，并辅以弧度，构成新的货架固定方式，同时也打通空间各处，令空间成为连接贯通的整体，让顾客在店内的行走几乎无阻碍，畅通且自在。

[2] 突出曲线的线条美

货架高低起伏，曲折通幽，或是波浪形，或是圆弧扇形，主次分明，在处处连接当中赋予空间和服装曲线美，形成服装展台的既视感。

[3] 利用造型，巧设陈列台

设计师充分利用空间，即便是鱼眼造型的落地点也不浪费，以之为柱为桌角，导出小小的桌面，无论是展示服装，还是单纯作为一个视觉补充和美观呈现，与四周的货架相匹配，都能起到让顾客稍作停留的实际功用。

[4] 在视觉入口，搭配经典色

落地玻璃隔绝店内与店外，此处也是大部分消费者看到店铺的第一处，因而设计师安放了4位模特。模特的中长款服装构成了黑与红的经典配色，同时透过玻璃还能观察到室内情况，能很好地起到吸引消费者进店的作用。

灯光设计

Lighting Design

[1] 善用光与影，突出品牌特色

在后现代工业风的环境中，灯光与空间结构的影相结合，营造出暗黑的氛围，体现GYWJ品牌暗黑的文化主题，更表现出品牌在暗黑色系下寻找自我、追随自我、突破个性的特点。

[2] 吸引顾客，让灯光成为另一种销售

空间里仅设置少量的必要光源，且是嵌入天花板部位的射灯，使得空间没有强硬的炫目之感，而是适应眼睛舒适度的光效。所展示的服装处在较为均匀同时光影突出的位置上，有一种十分巧妙的夺人眼球的作用，如此一来，灯光俨然充当了“不会说话的

[3] 光与结构

顶面和侧面结构在光的照射下，有了更丰富的条理，分明的亮度与暗度又让结构更显大气。

[4] 光与细节

在展销空间里，光的亮度需要恰到好处，光所带出的影子也需认真把握。在这个空间里，影子上下界线明显，其中服装的影子为竖向，小而聚拢，一方面增添店内层次，另一方面服装与地面的交织也不会给空间带来杂乱无章的困扰。

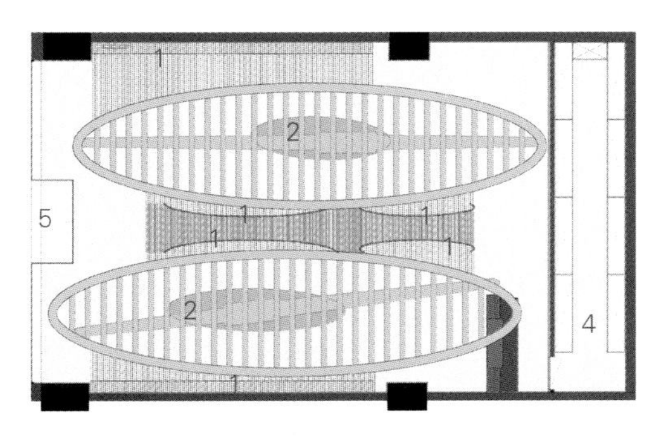

Overall Structural Analysis

1 Sales Area 销售区
2 Display Table 展示桌
3 Checkout Counter 收银台
4 Storage 储藏室
5 Entrance 入口

THE ORNATE AREA OF LUXURY CLOTHING

奢侈品服饰的华丽据点

MONCLER FLAGSHIP

The down jacket brand—Moncler, which was founded in 1952 in the French mountain village of Monestier-de-Clermont and is now based in Milan, has been personally led by the chairman and managing director Remo Ruffini. The new New York flagship store has mainline for men and women, Grenoble special mountain series, Moncler Enfant and other accessories. This store was designed by the French architect "Gilles & Boissier".

羽绒服品牌——盟可睐于1952年在法国小镇Monestier de Clermont成立，现在位于米兰，由董事长兼总经理Remo Ruffini亲自领导。全新的纽约旗舰店拥有男装和女装的主线系列、Grenoble特别登山系列、Moncler Enfant以及所有配饰产品等。这家店由法国建筑师双人组“Gilles & Boissier”操刀设计。

项目信息 PROJECT INFORMATION

Project Name | Moncler Flagship

Design Company | Gilles & Boissier

Designer | Gilles & Boissier

Project Location | New York, US

Area | 603 m²

Photographer | Eric Laignel

ELIASSON

空间创意
Space Originality

As with the most glamorous of apartments, there are acres of marble. Slabs for the floors and walls are caramel-streaked chocolate, gorgeously figured black, or snowy white. The designer gave a catwalk-esque vibe to the main corridor, paved with graphic black-and-white chevrons. It runs from front to back and terminates with the grandest of oversize gestures, a large bronze of a man's head.

Beyond the foyer, the two-level, 6,500-square-foot space continues with an impressively long and wide corridor passing an enfilade of salons with different themes.

1/2 | 3

就像最迷人的公寓一样，这里铺设了大量的大理石。地板和墙壁的石板是焦糖条纹巧克力色，或华丽的黑色，或雪白。设计师用黑白相间的图案铺成主走廊，创造出一种走秀T台的感觉。走廊从前到后延伸，并以最宏伟的装饰——一个巨大的男人头铜像结束。

除了大厅外，二楼约603.8㎡的空间同样延续着这令人印象深刻的长而宽阔的走廊，走廊串联起一系列不同主题的沙龙。

SCULPTURE

室内陈设
Indoor Furnishings

[1] 等候区

舒适的等候区是大型店面的设计重点之一。本案的等候区除了舒适的沙发、美观的茶几之外，还摆放了许多杂志、图书，让顾客打发等候时间。这里运用关联陈列，舒适的等候区四周摆放服装之外，还有鞋、包等，这在无形中增加了销售机会。

[2] 空间修饰

设计师在室内摆放了许多雕塑、书籍与挂画，增加了空间的文化底蕴与艺术品位，更彰显了品牌地位，给消费者一种高雅的空间体验。

SCULPTURE
MICHELANGELO
Redouté Most Beautiful Flowers
TERRY
RICHARDSON
VOGUE

DALÍ
DALÍ I
DALÍ II
SEBASTIÃO SALGADO
GENESI
TASCHEN
A RETROSPECTIVE
teNeues
VELE CLASSICHE
ALPI

[3] 服装的对比陈列

对比陈列法的优势就在于突出渲染商品，烘托空间气氛，给顾客以深刻的印象。蓝色的服装在黑色的空间中因光线的着重表现而显得十分亮眼，而热烈的红色地毯更是反衬了服装的存在。

[4] 产品陈列

空间大部分陈列采用格子式布局，整齐、清晰的陈展给客人明确的参观视角，在满足审美的同时，争取达到促销的最终目的。

灯光设计

Lighting Design

[1] 灯光改变空间层次

空间布局除了本身需要具有层进之外，还离不开灯光的辅助。在吊顶内设置隐藏光源，从而达到突出层次感的作用。

[2] 灯光构成交通动线

廊道顶灯采用小型射灯规律排列，成为动线指引的一部分。

1 | 2
4 | 3

[3] 空间配色

空间许多地方都用了美国橡木做修饰。炭黑色的空间中加入大片热烈的红色，这一经典撞色在白色灯光中暴露出来，创造出深邃神秘而热切的空间气氛。

[4] 灯光与艺术装置

通常在门店出入口的灯光效果要强些，这样不但有助于顾客辨认门店，还能保证安全性。本案入口处的艺术装置在灯光的配合下尽显弧度，增强空间的视觉张力，提高了专卖店的辨识度。

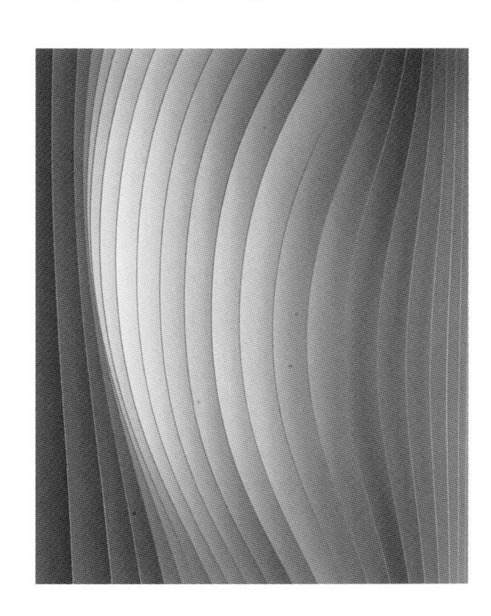

[5] 强化中岛部分，提升空间连贯度

在展柜处增加光源是店铺灯光设计的惯用手法。通常是增加展柜的亮度，或者增加展柜中产品的聚光灯效，以达到突出主体的目的。

[6] 暖光灯效

就像一座宏伟的巴黎公寓，它的序幕始于一个合适的门厅。顾客通过路易斯风格的油浸青铜拉手打开了高高的玻璃前门。空间两侧的墙上安装着高大、弯曲的百叶窗，阳极氧化的黄金色灯具以背光的方式辐射出一种阳光般的温暖。

[7] 灯光下的大理石

楼道与整个收银区铺满了大理石，白色洗墙灯赋予石材花纹以动感，使得墙面仿佛是流动的液体，给顾客一个非常独特且印象深刻的空间体验。

REPRODUCE HISTORY AND REAPPEAR MODERNITY

重塑历史，再现摩登

|OIAM|

OIAM Boutique is located in the center of the French Concession on Wukang Road, Shanghai. The road flanked by chinar is permeated with the tranquil and elegant French life atmosphere. Mix and match is her personality. The three-story modern-style house of about 600 square meters fulfills the function of shopping, recreation and workshop.

OIAM时尚买手店坐落于上海武康路法租界中心地段，梧桐婆娑之下是静谧优雅的法兰西生活气息。混搭是她的个性，购物、休闲、作坊等功能囊括在近600㎡的三层洋房里。

扫码查看电子书

项目信息
PROJECT INFORMATION

项目名称 | OIAM时尚买手店
设计公司 | NONG Studio
主创设计师 | 汪昶行
设计团队 | 朱勤跃、魏林、焦凤林、刘雨立
项目地点 | 上海
项目面积 | 600㎡
主要材料 | 大理石、不锈钢、黄铜、青砖、手绘天花、定制地毯、垂直绿化、水磨石、定制墙纸、定制窗帘、毛毡、做旧木地板等

SWAROVSKI
FURLA
DIESEL
BURBERRY LONDON
SHIATZY CHEN

空间创意 Space Originality

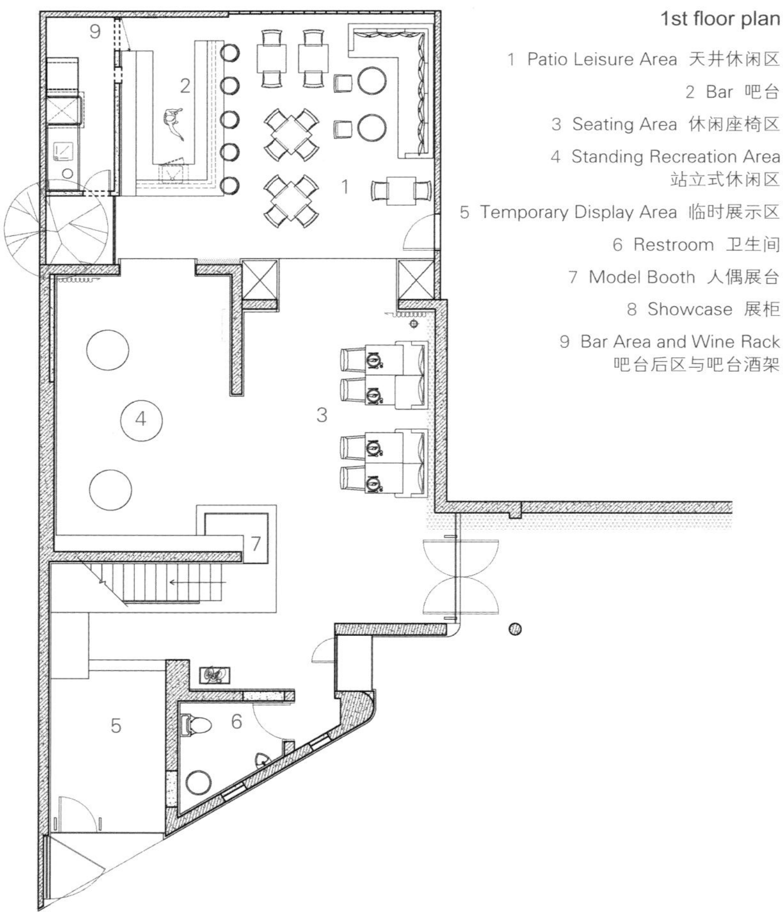

1st floor plan

1 Patio Leisure Area 天井休闲区
2 Bar 吧台
3 Seating Area 休闲座椅区
4 Standing Recreation Area 站立式休闲区
5 Temporary Display Area 临时展示区
6 Restroom 卫生间
7 Model Booth 人偶展台
8 Showcase 展柜
9 Bar Area and Wine Rack 吧台后区与吧台酒架

| 一层 |
THE FIRST FLOOR

On the whole, the theme of the first floor is bold and avant-garde. Even its details manifest modern and fashion, such as the branch-type chandelier behind the iron fence, the Dali red lips Sofa, the Philippe Starck golden gun lamp and the most surprising touch of the vault door for the bathroom. However, everything is hidden behind the greens. Only the neon light will meticulously outline the enchanting urban life silhouette in the evening.

一楼的调性是张扬前卫——铁栅栏后的枝型吊灯、弧形玻璃窗里的达丽红唇沙发、倒挂的模特、角落里Philippe Starck的金色枪灯，另外还出人意料地将厕所门制作成金库门，连细节都毫不造作地张扬着摩登时尚，但一切都隐在一片绿丛之后，只有霓虹灯会在傍晚细腻地勾勒出风情万种的都市生活剪影。

室内陈设
Indoor Furnishings

[1] 夸张的个性

混搭是她的调性，游走于公共空间、私密、半私密之间，外观的内敛复古与室内的摩登夸张强烈地冲击着顾客的视觉体验。

[2] 独特的展示

一楼的焦点莫过于倒挂的人体模特，能给人强烈的视觉冲击感。零售空间要给人新颖感，且印象深刻，就需要在一些细节上出人意料。

VOGUE
BALLY

VALENTINO
DKNY

灯光设计

Lighting Design

[1] **丛林般幽谧**

设计师在宽敞的一楼空间做了静谧浪漫的灯光氛围，让人联想到《绿野仙踪》丛林里的感觉。用亚克力打造的店铺LOGO通过灯带的修饰变得独特而时尚。

[2] **装饰与灯光的配合**

唯美的白色玫瑰墙在顶部射灯的配合下有羽毛般的纯洁轻盈感；工业风造型的金属吊灯在大型试衣镜的搭配中增加了空间的延伸性。对于不同空间视觉效果的思考是设计师必须进行的修炼。

| 二层 |

THE SECOND FLOOR

On the second floor, its function wanders between the display and living room. The soft white window gauze builds a comfortable and gentle shopping experience. You can flip through VOGUE sitting in the Wegner shell chair or have small talks, slumping in the peacock blue vintage sofa in the VIP room. The golden mat glass behind the white window gauze can change the interior atmosphere according to the season.

二楼的调性游走在展厅和客厅的边缘，柔和的白色窗纱创造出温柔舒适的购物体验，你可以在Wegner的贝壳椅里翻翻时尚杂志，也可以约在VIP室孔雀蓝的复古沙发里闲聊八卦。隐约在白色窗纱后的金色磨砂玻璃可以根据季节不同，改变整体室内场景氛围。

室内陈设
Indoor Furnishings

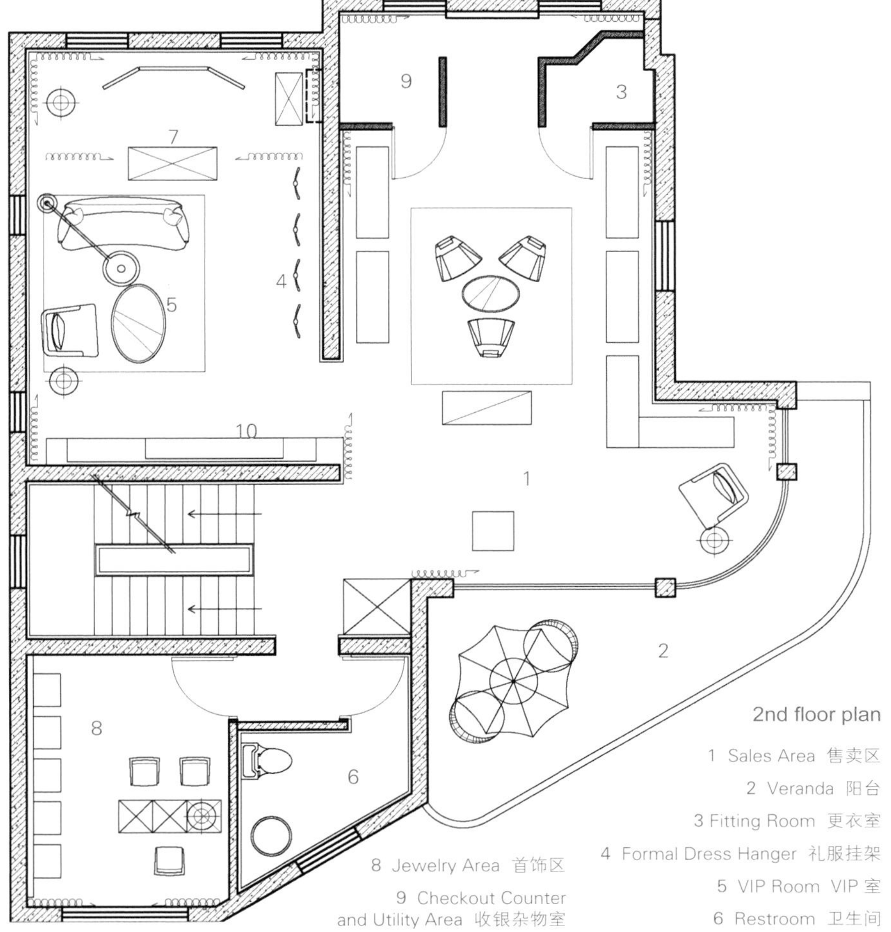

2nd floor plan

1 Sales Area 售卖区
2 Veranda 阳台
3 Fitting Room 更衣室
4 Formal Dress Hanger 礼服挂架
5 VIP Room VIP 室
6 Restroom 卫生间
7 VIP Fitting Room VIP 更衣区
8 Jewelry Area 首饰区
9 Checkout Counter and Utility Area 收银杂物室
10 Wood Furnishing Display Case 木饰面展示柜

[1] 优雅的二楼

二楼的设计带着复古的柔情，天花板上欧洲古典风格的画像更增添了空间复古的时尚情调。服装展示不以数量为先，隔空两列排开，给人清晰舒爽的视觉感受。

[2] 服装搭配

在服装展示区采用关联陈列的方式，专门挑选了能与所展示服装相搭配的女鞋，方便顾客在选购过程中找到对的感觉，提高服务质量的同时增加店铺销量。

|三层|
THE THIRD FLOOR

3rd floor plan

1 Clothing LAB 服装 LAB
2 Balcony 阳台
3 Office 办公室
4 Communication Area 洽谈间
5 Storage/Clothing Shooting Area 仓库 / 服装摄影区
6 Restroom 卫生间

楼梯立面图

1 Display Window 橱窗
2 GL Glass Partition GL 玻璃隔断
3 Glass Curtain Roof 玻璃幕墙顶
4 Wall Lamp 壁灯

AFTER RENOVATION

PLAN ANALYSIS 1F

1 Showcase Space 陈列区
2 Staircase 楼梯
3 Display Area 展示区
4 Windows Display 橱窗展示
5 Sun Glass Room 太阳眼镜区
6 Toilet 洗手间
7 Front Garden 前庭花园

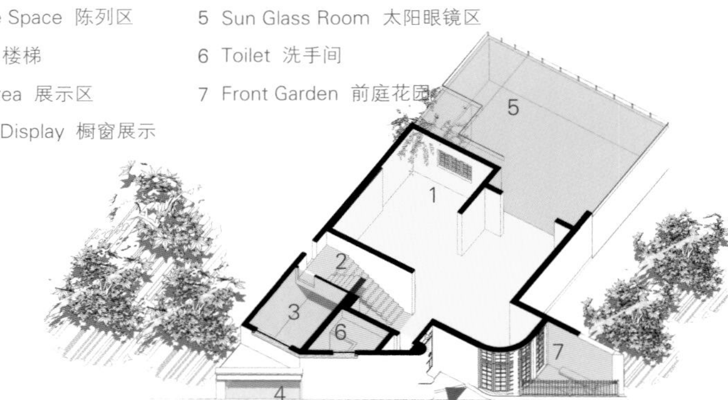

DESIGN CHALLENGE

1 garden space transformed into semi-open space
花园区域改造成半开放区

2 lacking of nature light
缺少自然光线

3 too low to walk through
太矮难以通过

PLAN ANALYSIS 2F

1 VIP Room VIP 室
2 Staircase 楼梯
3 Jewelry Room 珠宝室
4 Showcase Space 陈列区
5 Toilet 洗手间
6 Balcony 阳台

EMS SOLVING

1 open the ceilingof staircase to bring more natural light indoor
开放楼梯的天花将自然光线引进室内

2 sunglass room for vip lounge
太阳眼镜室作为 VIP 休息室

3 tear down floor to emphasize the transition between spaces and scales to create a rhythm for the entrance
拆除地板，强调空间和比例之间的转换，为入口创造一种节奏感

3F PLAN ANALYSIS

1 Office 办公室
2 Staircase 楼梯
3 Storage 储藏室
4 Toilet 洗手间
5 Workshop 工作坊
6 Balcony 阳台

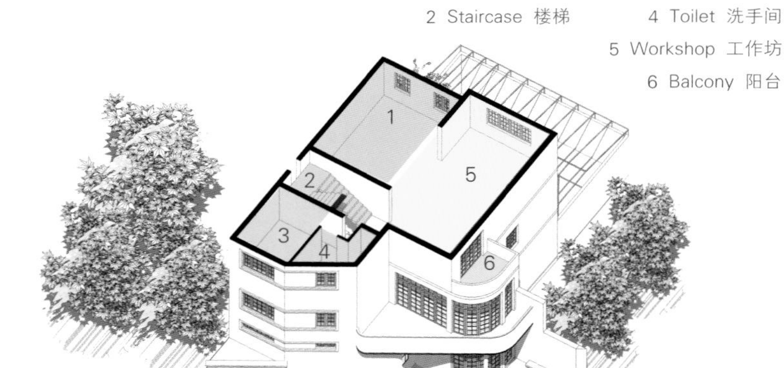

ROOFTOP PLAN ANALYSIS

1 Rooftop Garden 楼顶花园

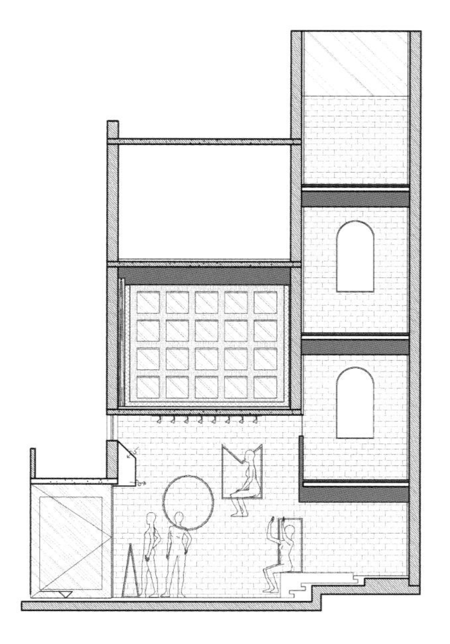
楼梯立面图 & 南侧房间立面图 1

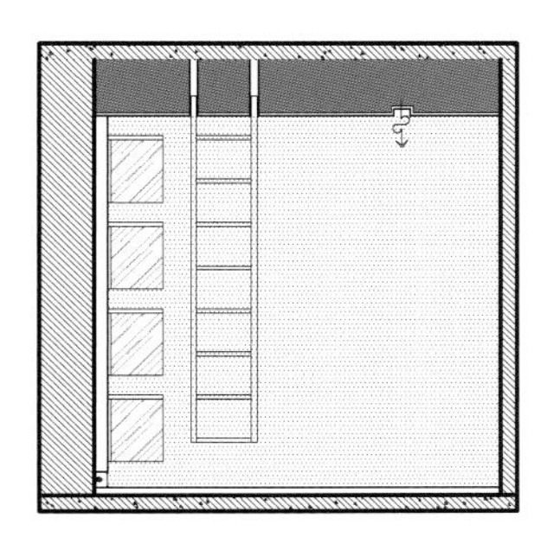
南侧房间立面图 2

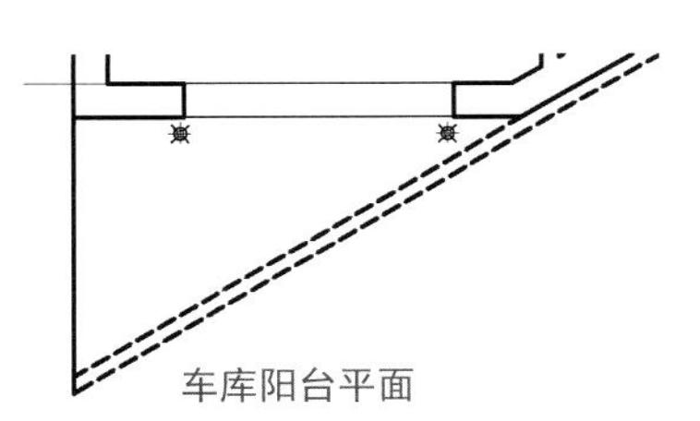
车库阳台平面

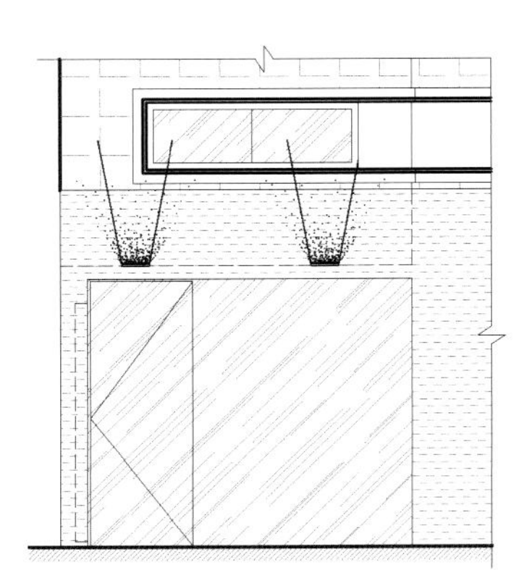
车库阳台外立面

NIKE SHANGHAI MARATHON EXPO 2017

2017上海国际马拉松·耐克

品牌文化 Brand Culture

|NIKE|

Our mission is bringing inspiration and innovation to every athlete in the world. If you have a body, you are an athlete. The sustainable innovation is a catalyst for revolutionizing the way we do business and an opportunity that's been integrated across our business in policies, processes and products. We are innovating solutions that benefit the athletes, the company and the world.

我们的使命是给世界上每一位运动员带来灵感和创新，如果你有一个身体，你就是一名运动员。可持续创新是使我们的经营方式发生革命性变化的催化剂，也是一个将政策、流程和产品整合到业务中的机会。我们正在创新有利于运动员、公司，甚至是利于世界的运动产品。

项目信息
PROJECT INFORMATION

项目名称 | 耐克 2017 上海国际马拉松体育博览会

设计公司 | 协调亚洲 COORDINATION ASIA

项目地点 | 上海

项目面积 | 1600㎡

扫码查看电子书

NIKE+ RUN

WELCOME TO NIKE+ RUN CLUB
克上马配速员

NRC
NIKE+ RUN CLUB

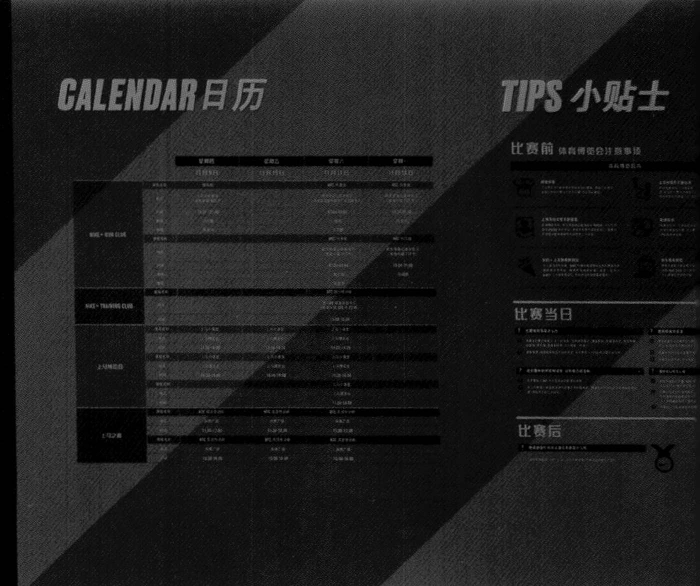
CALENDAR日历
TIPS 小贴士
比赛前
比赛当日
比赛后

空间创意

Space Originality

The Nike Shanghai Marathon Expo 2017 brand space is all about getting faster and breaking records.A large overhead animated LED-installation displays countless target times and these are runners' plans and expectations that they can reach or break their personal bests. The center of the 1,600-square-meters space marks a trial-run-cage where runners can get tips and tricks of experienced coaches and try Nike's Zoom Fly running shoes. Pacers who will run the marathon in a certain time introduce themselves and invite those runners to follow. A highlight is an-18-meter-long wall lists all 38,000 names of those who will participate in the 2017 Shanghai Marathon.

JUST DO IT
SHANG马HAI
2017
5K
10K
42K

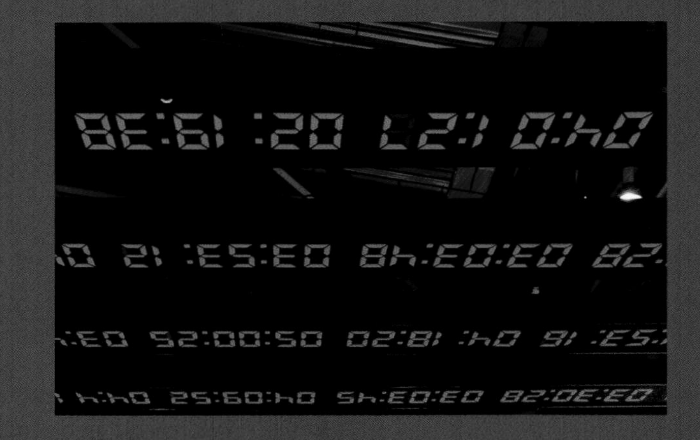

耐克2017上海国际马拉松体博会品牌空间致力于打造一个更快速、破纪录的氛围。场地上方悬挂的大型动画LED装置展示着数不尽的目标成绩，这些成绩是跑者对超越自己、突破极限的计划与期望。1600㎡空间的中央是铁丝网打造的试跑专用平台，在这一区域，跑者能得到更多来自经验丰富的跑步教练的小贴士与技巧，同时可以试穿耐克的Zoom Fly跑鞋。空间亮点则是长为18m，有着2017年上海国际马拉松38,000名参与上海马拉松长跑者名字的姓名墙。

TRIAL RUN
NRC
NIKE+ RUN CLUB
NRC
NIKE+ RUN CLUB
NIKE+ RUN CLUB

RUN CLUB
你的完美
跑步训练指档
CALENDAR日历
TIPS小贴士
试
TRIA

RUN
NIKE ZOOM FLY
突破挑战

室内陈设
Indoor Furnishings

1 | 2
3

[1] 视觉感受

粗犷的色彩方案由黑、白、红三色组合，强烈的视觉对比，营造具有沉浸感、热切的氛围，与积极动感的背景音搭配，通过现场众多的扬声器在空间中不断激荡回响。

[2] 服装摆放

设计师非常大胆地在黑色展示区展示素色的服饰，通过灯光将服装从空间中突显出来；红色展示区摆放红色的服装，整齐而显眼。室内采用斜角式布局，为空间增加延伸感，让内部布局变化具有空间性；另外再综合使用格子式布局给人秩序感。室内跑步姿势的模特为空间增加许多富有朝气的动感。

[3] 简洁扼要

鞋品展示区十分简洁，通过灯管由外向内、由下至上照射来体现跑鞋的形象。

JUST D
NIKE+ RUN CL

NRC
NIKE+ RUN CLUB
MEN'S
NIKE+ RUN CLUB

空间装饰

两个大型LED墙面展示了耐克最新推广短片《有种快叫上海》，镜头追随一位年轻运动女郎冒险般地穿梭奔跑于上海的大街小巷，追求更快的速度和更大胆的探索。影片由尼古拉斯·温丁·雷弗恩（《亡命驾驶》导演）执导，Wieden + Kennedy上海公司制作。

天花板上是许多定格的跑者的目标时长，非常契合上海马拉松的主题。

05:30:34
04:48:29
03:04:56
88:88:88
88:88:88
03:05:35
04:35:39
02:39:48
NIKE
NIKEAIRMAX

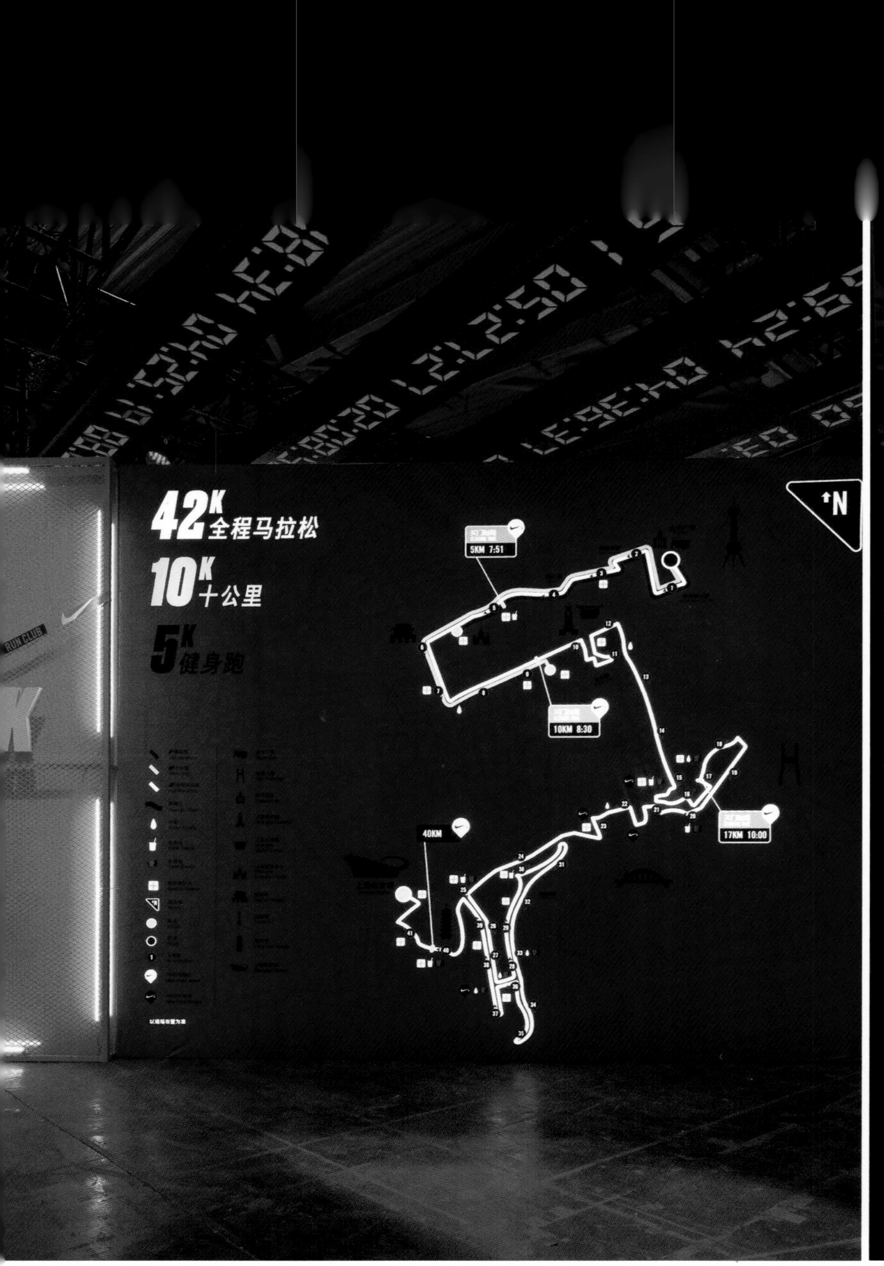

灯光设计

Lighting Design

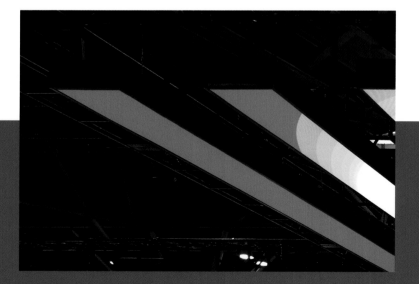

[1] 综合材料

将铁丝网、粗犷的水泥地面等工业元素与时尚的平面图文、都市设计元素诸如外露的霓虹灯管、动感的媒体装置相结合，设计团队通过这些来呈现具有震撼效果的品牌环境。

[2] 灯管的使用

大量使用灯管、灯带，在黑色的空间中给人清晰酷炫的线条感，给空间炫彩红色的工业风气质，制造出不一样的粗犷大气的环境。天花板的条状LED屏幕除了显示时间，还不时变化色彩和内容，给空间赋予律动感。

SHANGHAI
SHANGHAI
SHANGHAI
上海

THE STREET STORE, THE CONTRASTING FASHION
街边的店铺，对比的时尚

NGUYEN HOANG TU

Nguyen Hoang Tu is a young fashion designer who has shaped his direction as "an independent artist"—showing through his collections that have been performed domestically and internationally. Nguyen Hoang Tu flagship store will be the place for selling the limited conceptual fashion products, dedicated to women and honoring the value of silk as well as the handicraft details.

Nguyen Hoang Tu是一位年轻的时装设计师，他为自己树立了“独立艺术家”的发展方向——已经在国内外的展览上展出过他的作品。这家旗舰店将是有限概念时尚产品的销售场所，专为女性客户服务，并尊重丝绸和手工艺品的价值。

扫码查看电子书

项目信息
PROJECT INFORMATION

Project Name | Nguyen Hoang Tu flagship store

Design Company | Red 5 Studio

Designer | Lai Chinh Truc

Project Location | Hochiminh City, Vietnam

Area | 60 ㎡

Photographer | Quang Tran

空间创意 Space Originality

Right next door is also a very eye-catching fashion shop. After scrutinizing the site and assessing the situation, the designer knew that this project needs a lot of effort to change the facade entirely as well as the interior space to achieve the intentions of the architect and the owner. “Contrast” is the decisive method: the facade contrasts with the adjacent stalls, and the interior space contrasts with the display products.

正好隔壁也是一家引人注目的时尚店。在这种情况下，经过仔细检查和评估，设计师认为这个项目需要大量的工程来完全改变立面和内部空间，以达到最终的设计意图。“对比”是决定性的方法：正面与相邻店铺形成对比、内部空间与展示产品形成鲜明对比……

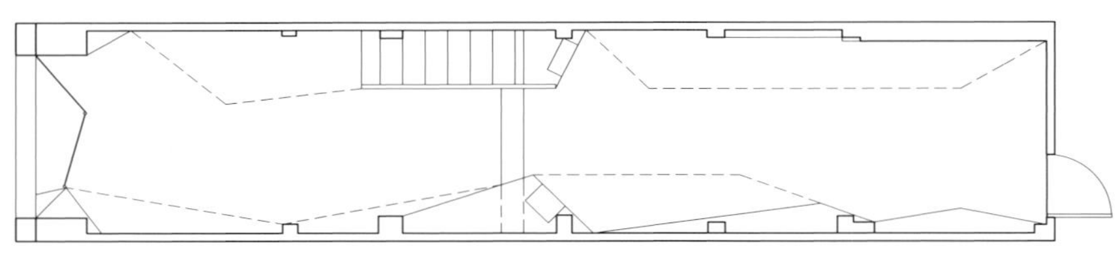

GROUND FLOOR PLAN

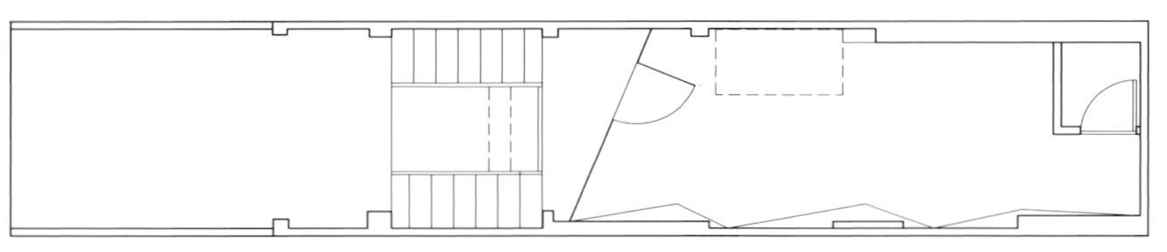

1ST FLOOR PLAN

室内陈设
Indoor Furnishings

[1] 展示的意义

店铺方形的门框被做成裸露混凝土的样式，而门则向内形成一个倾斜的角度。当门打开的时候，我们会第一眼注意到展示在门口的产品，这个区域是突出展现店铺当时最想展示的作品的地方，这体现了店铺产品展示的意义。

[2] 室内空间的“对比”

室内空间符合“对比”的主题原则。材料上，如水泥、玻璃和光与立方体、锯齿形的配合，创造了一个令人印象深刻的展览空间。白漆环氧地板创造出了跑道的感觉；上层阁楼有意保留部分旧天花板和胶泥，所有这些都适合创造一个整体，代表了“对比”的语言。

灯光设计
Lighting Design

[1] 奇幻的镜面

设计师在镜面底边设置白光洗墙灯，给顾客提供了一条清晰的交通动线。三种反射镜具有扩大空间视觉效果和创建虚拟感的作用，并让灯光达到事半功倍的效果。

[2] 层次

整个空间以白色为底，向上做深灰色空间，两者形成一个对比，灯光也是如此，以地板为起点向上减弱，从明暗对比中显现出空间的层次。

PLAINNESS BUT EXQUISITENESS AND ELEGANCE

质似朴，实则细腻又优雅

| MINZE-STYLE |

MINZE-STYLE has always been a leading brand in the woman's fashion industry of China, and its fashion is personalized but restrained, paying attention to the fashion inheriting the classic and the integrity of classics and soul. From "personal tailor, beauty theory, social brand and private brand", it inherits the traditional design style of Italian fashion, incorporates with the physical beauty of the oriental ladies in metropolis and the various favorite elements around the world, meets the needs of the high-end female customers for the quality of life and brings the customers an exquisite and elegant experience.

MINZE-STYLE名师路一直是中国时尚潮流女装行业的领先性品牌，时装个性而不张扬，讲究时尚传承经典以及经典与灵魂的结合。其从“私人定制、美丽学说、社交品牌、自有品牌”四个方面出发，继承着意大利时尚传统魅力的设计风格，结合东方大都市女性的形体美以及世界各地不同的流行元素，满足高端女性客群对品质生活的需求，为客户带来精致与优雅的体验。

项目信息

PROJECT INFORMATION

扫码查看电子书

摄 影 师 | 吴永长

主要材料 | 素水泥、钢板、金属网等

项目面积 | 1200㎡

项目地点 | 福建福州

设计团队 | 刘任玉、肖达达

设计主创 | 何华武

设计公司 | VHD维度华伍德

项目名称 | MINZE-STYLE名师汇

The interior space is supposed to present the charm of the primitive and wild nature, and the steel cladding is used to bring out a sense of reality. This kind of straight framework materials, the directness and simplicity produced by space and a sense of power or lightness by the oversized steel plates, all of these create a continuity relationship between the whole space and the original architecture concerning time and space.

Taking advantage of the flexible nature of "arch", designers create different spaces so as to avoid the similar experiences. This is a process of recalling the environment of the nearby park which establishes a dialogue between the building scale and its space again. The simpler the shape is, the more profound impression it offers. Endowed with a gentle curve, the arch presents beauties from different angles. The continuous curves of arches create the enclosed areas. In this way, the interior and exterior spaces are able to maintain the spatial continuity, and this idea helps space unconventional and explores a new order of the traditional space types. "Arches" play an important role in the traditional architecture and have constantly been enriched throughout the history. Designers are trying to change the conventional "arch" so that they can achieve a breakthrough inhomogenous and inflexible modern spaces, by which a natural power is acquired and is turned into the spatial orders. The strong spatial aesthetics reinforces the significance of brand.

空间创意

Space Originality

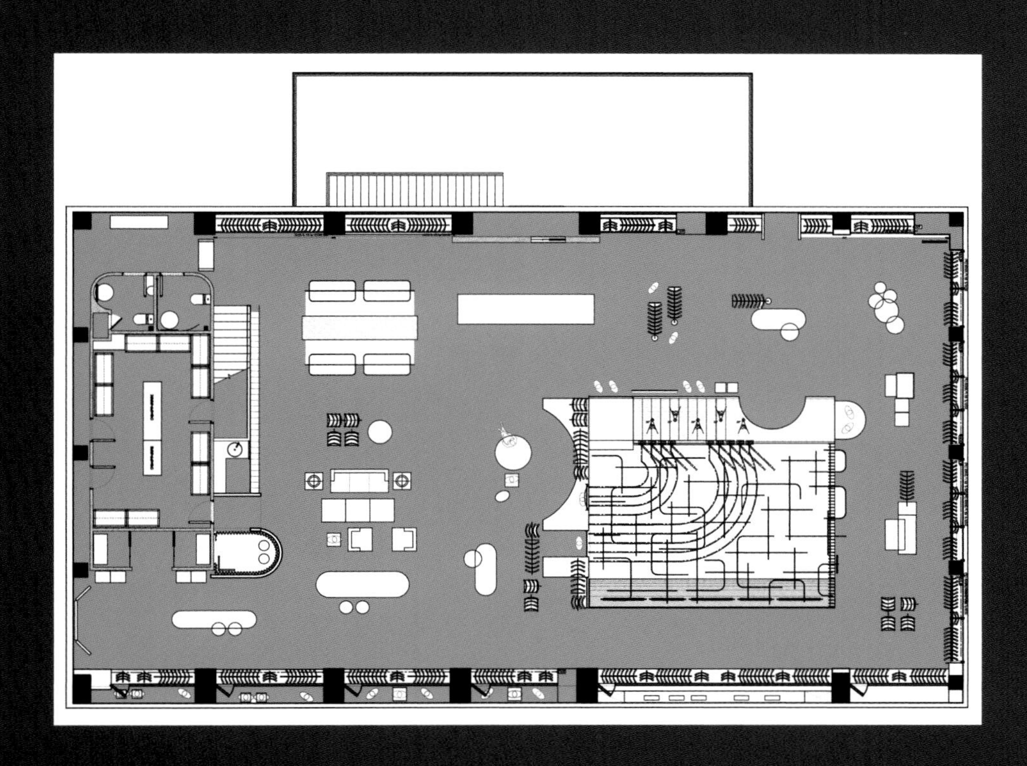

本案设计是一个对附近公园环境回忆的过程，再一次让建筑的体量与空间对话。设计师希望内部空间呈现一种原始野性的魅力，故而采用钢制的表皮带来一种现实感。这种直白式的架构材料，空间所形成的直接性与朴素性，加上大尺度的钢板落地产生的力量感或轻盈感，使整个空间与原有场地建筑取得一种时间与空间的接续关系。

“拱”在传统建筑中扮演重要角色，并且在漫长的历史中不断丰富。设计师试图转化传统“拱”，让“拱”在均质呆板的现代空间里制造新的突破，获得自然能量并把它转化为空间秩序。利用“拱”多变的空间特性，产生不同的空间，也表现了越是简单的形体逻辑，越能给人带来深刻的印象。弧拱曲线柔和，呈现不同角度的美感，其连续的曲线把空间构造成围合区域，使得内部和外部保持着连续性，也使空间不拘泥于常规，探索传统类型空间的新秩序，而浓烈的空间美感也显示了该品牌的重要意义。

室内陈设
Indoor Furnishings

1 | 2
3

[1] 依商业需求，灵活设计

设计师根据品牌的需求，进行模块化设计，使品牌和店内的不同功能具有最大的灵活性。

[2] 以拱造形，去除多余装饰

弧拱同空间一、二层相连，设计师将不同的“拱”垂直交叠，创造不同的空间与体验；利用“拱”的多变空间特性，打造具有现代感的室内空间，创建出有趣的空间状态，给人们提供强烈的视觉效果，引起顾客的好奇心。

[3] 木屏风与货架相结合，渲染服装销售的多样性

和一般常规服装店陈列的概念不同，这家服装店强调建筑化的零售模式。纤细的木屏风结合货架矗立在边界处，渲染了空间和室内饰面的丰富多样性，与此同时建筑元素显露材质原始的质感，烘托服装的品质，为整个店铺灌注纯粹的气氛，隐喻品牌的精神。

[4] 让时装模特引起消费者的注意

模特作为不可或缺的道具，不仅填补了空间的空白，而且是空间留给消费者最深刻的印象之一。在一大一小拱形门的宽阔区域，在二层楼梯的尽头，在窗边，在桌旁，这些模特顾盼生姿、婀娜轻盈，消费者一抬眼，便能收获一份美妙。

灯光设计
Lighting Design

[1] 引入自然光，贴近城市生活

设计师为空间预留了天井，天井楼梯释放出不同的光影效果，使整个空间与建筑取得一种时间与空间的对话，在实现品牌空间本质功能的同时，还续写着城市环境的延续与变迁。

[2] 局部投射，大部柔和

基于弧拱结构形成的展厅情况，设计师多用联排式的射灯，既使服装展示区亮堂，让服装本身得到顾客的重视，又能在成拱的区域形成细节亮点；加之朴素而雅致的刚面与玻璃，在整体柔和的视觉感受中，更突出建筑的简洁与现代美感。

1 | 2
3 4 5

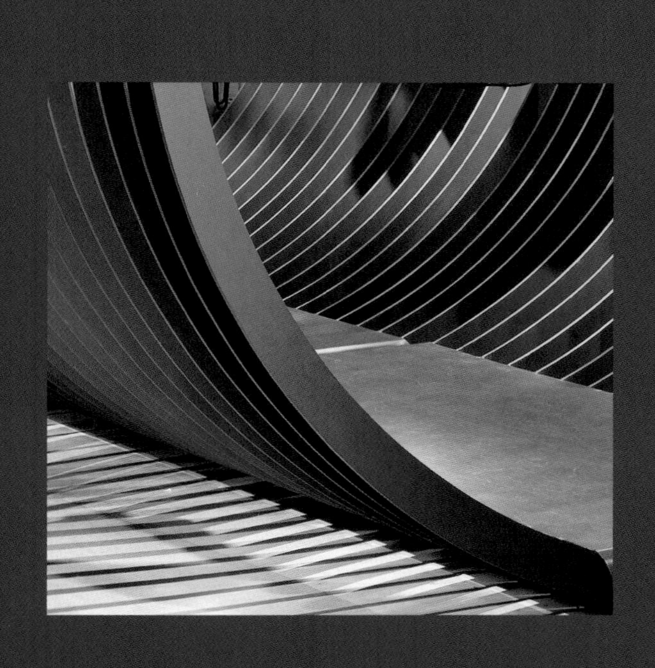

[3] 光与地面

U字形的展示区间，光影之下，黑白分明、隐隐约约，而地面如水，倒映造型，显现空间的多变。

ABANDON INEFFICIENCY, EXPLORE THE INTEREST OF STRUCTURE

抛除低效，探寻结构趣味

GUAN × K-HOUSE

The store is the concept store of the cooperation between the Guan brand and K-HOUSE, both of which are relatively young menswear brands. The brand Guan was founded in 2015, and the designer interprets the new understanding of clothing from a more subjective perspective for the middle-class men who are between 25 to 40 years old with the same clothing preferences and finds the innovative, personalized, textured and cost-effective clothing to them. The word "Guan" is taken from "clairvoyance", which means that wisdom makes people consciousness. The brand tries to find clear and sincere self-expression, hopes to use the independent thinking and the original spirit to create a new vision and make clothing become an art.

项目信息
PROJECT INFORMATION

项目名称｜灌×K-HOUSE
设计公司｜叶槟室内设计工作室
设 计 师｜叶槟
项目地点｜浙江杭州
项目面积｜40㎡
主要材料｜铜、铝、艺术漆等
摄 影 师｜刘宇杰

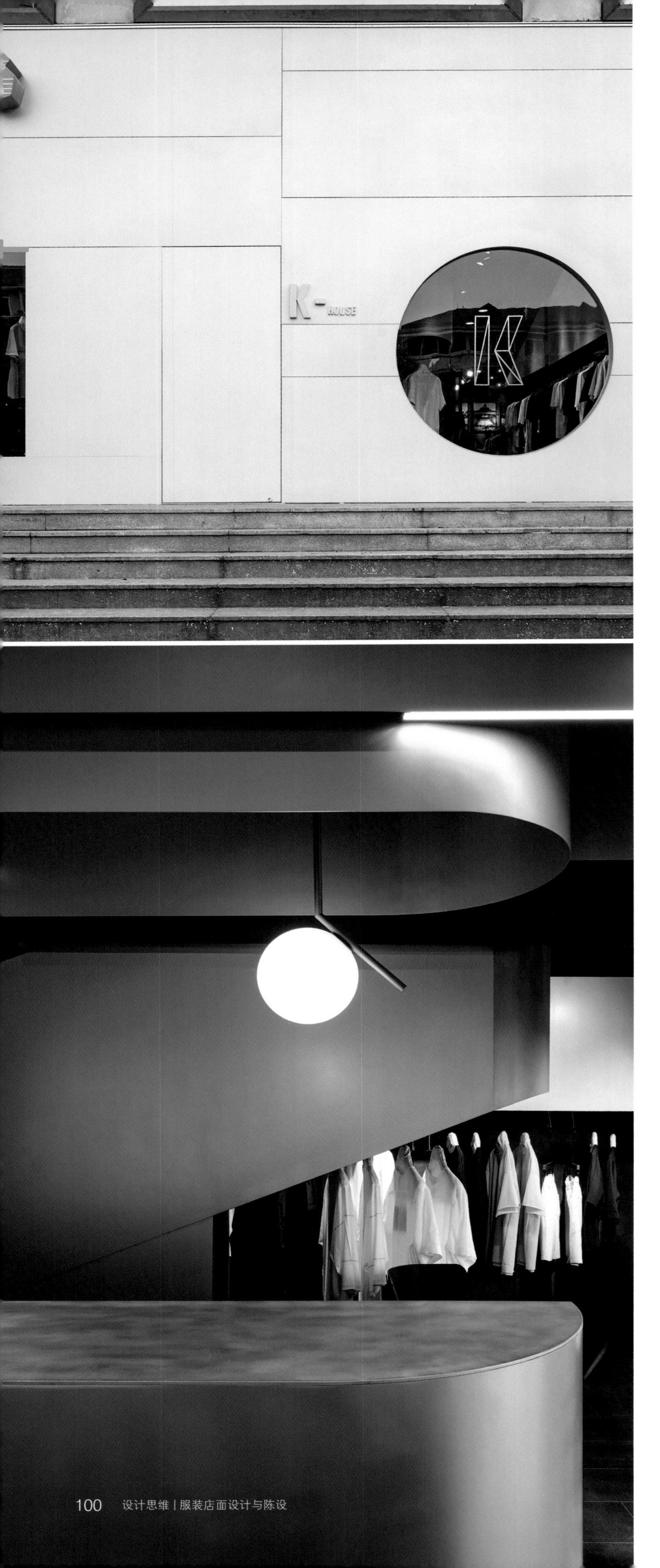

店铺是灌品牌与K-HOUSE合作的概念店，二者都是比较年轻的男装品牌。其中灌品牌成立于2015年，设计师以比较主观的视角诠释对服装新的理解，针对25～40岁有着相同穿衣喜好的中产阶级男性，为其开发原创、个性、质感、高性价比的服饰。“灌”字取自“醍醐灌顶”，寓意智慧使人觉悟。品牌试图寻找一种清晰而诚恳的自我表达，希望用独立的思考与原创精神创造新的视野，让穿衣成为一门艺术。

收银立面 1

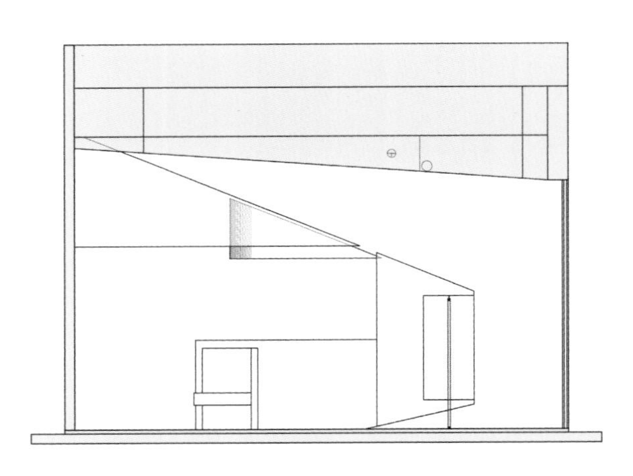

收银立面 2

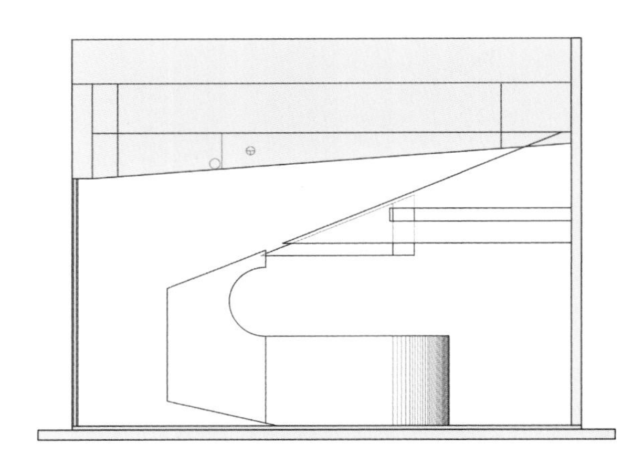

分析图 1

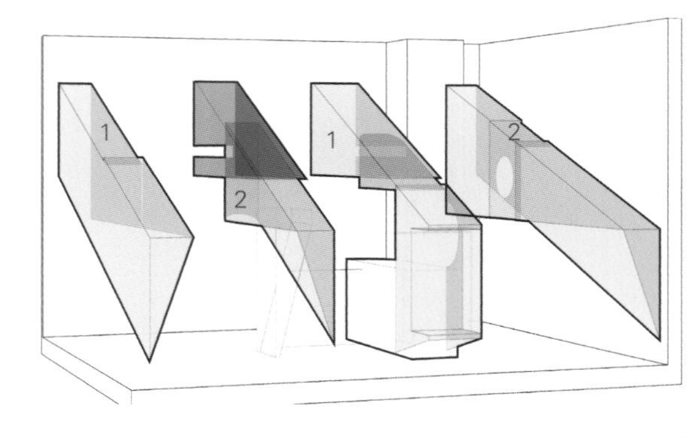

1 Grey Aluminium Board 灰色铝板
2 Bronze 铜板

空间创意
Space Originality

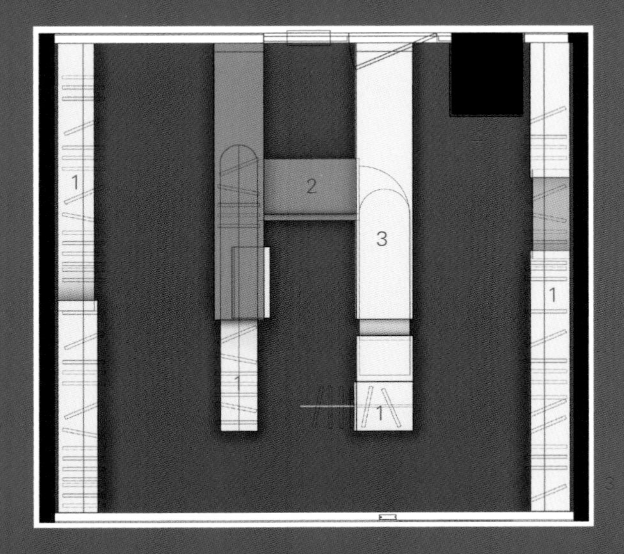

平面布置

1 Sales Area 销售区
2 Display Table 展台
3 Checkout Counter 收银台
4 Dressing Mirror 试衣镜

In a tiny space, the designer and the client challenge the new idea of throwing out all inefficiencies and strengthening display. The designer has creatively removed the fitting room in the traditional clothing commercial space. In addition to the deep thinking on the realistic aspect of the case, the designer focuses on the multi-dimensional thinking of the brand, and fully plays and extracts the confident control ability on products and design of Guan and K-HOUSE, so that put it in the design and display the core, and customers will trust the brand imperceptibly.

K-HOUSE

The main color tone of the shop is black, gray and gold, together with the material combination of bronze, aluminum and textured coating, so that space breathes a modern style. The facade uses gray aluminium boards to form the partition between indoor and outdoor. The striking golden LOGO is placed in an appropriate proportion and creates a mysterious atmosphere for the store. The hollowed-out of the front circle and rectangle forms a channel to peep inside, and many geometric shapes divide the interior to explore the subtle relationship between the plane and the solid in the vision. In fact, it is also a kind of thinking about the traditional clothing store mode, which is upgrading the new concept and new experience what brand conveys.

The cashier desk is hidden under the structure, and the vision and function are intertwined here, which originates from a kind of multidimensional commercial thinking—"how to guide the customer to acquire brand culture quickly in the commercial space display". The circular and semi-circle hollow out, corner details and so on, these details at each structural corner in space are handled seemly inadvertent, full of ingenuity and consistency.

在极小的空间里，设计师与委托方一起挑战了一个新想法，即抛除一切低效，强化货品陈列。设计师创意地去除了传统服装商业空间中的试衣间，除了对本案现实层面的深度思考，更多着眼于品牌的多角度思维，即将灌和K-house在产品与设计上自信的把控能力充分发挥并加以提取，使其居于设计和展示核心，潜移默化地使顾客将信任交付给品牌。

店铺的主要色彩基调为黑、灰、金，材质由铜板、铝板与肌理涂料等构成，趋于现代感。空间外立面用大面积拼接的灰色铝板构成内外的隔断，醒目的金色LOGO以适当的比例放置，神秘的未来感气质扑面而来。正面圆形与长方形的镂空镶入玻璃形成向内窥视的通道，多种几何形状对空间内部进行分割，探讨视觉上平面与立体之间的微妙关系，实则也是一种对传统服装店模式的思考，即升级品牌传达的新概念与新体验。

收银台藏于结构之下，视觉性与功能性在此交汇，这源于一种多维的商业思路——“如何引导顾客在商业空间展示中快速获得品牌文化”。圆形或半圆的镂空、转角细节等，空间里每个部分都经过了看似不经意的处理，充满巧思与连贯性的推敲。

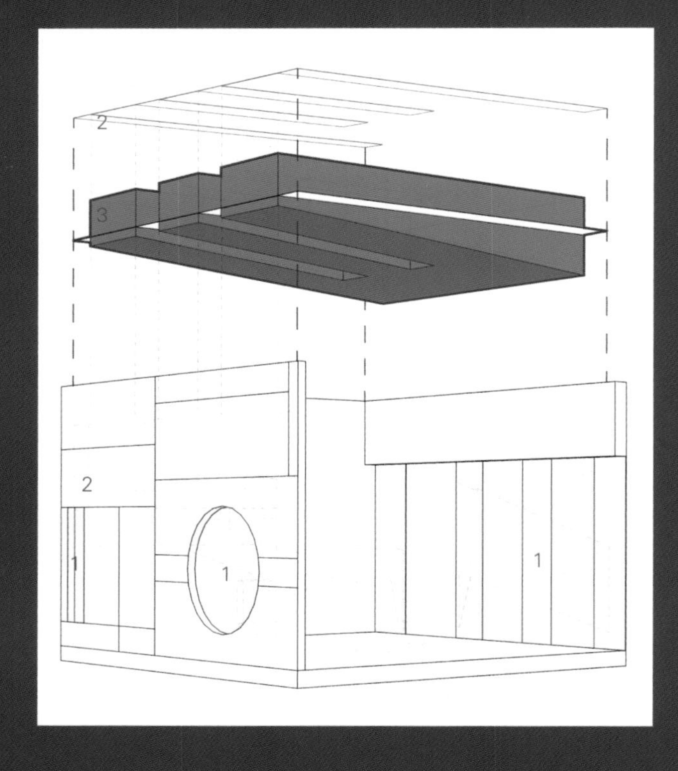

1 | 2

分析图 2

1. Glass 玻璃
2. Grey Aluminium Board 灰色铝板
3. Textured Coating 肌理涂料

[1] 路线明确，专注于陈列

设计师对空间场地进行了新鲜尝试和深度布局，去除了关联的中岛、试衣间，空间由此变得更适合进行纯粹的展示，一方面使进入空间时路线明晰，另一方面使空间更大化地利于陈列，为访客提供便捷式的服务。

[2] 从自身出发，诉求仪式感

多重交叠的空间排布和不同以往的导购线路相交融，服饰隐含在几何结构之中，富有概念感，不但带来不由自主的仪式感，也让人更专注于产品本身，营造出一种无杂质的购物体验。

[3] 在视觉上延展空间，凝聚向心力

不同颜色的服装在左右对称的空间中交错、交织，在全幅透光玻璃墙的作用下，对访客心理形成冲击效果，促进购买行为的完成。

灯光设计
Lighting Design

1
2 | 3

[1] 光与材质相呼应，营造高级感

外部光线从圆形窗户倾泻而入，进入色调深沉、冷静的内部空间，在对撞的材质、有趣的结构之间发生作用，形成微妙独特的空间变化。

K-
HOUSE

1 | 2
3

[2] 创意照明增添店铺趣味

主要的照明用灯和个性化的装饰用灯在黑、灰的大气空间里，如同宇宙间的月亮星辰，与空间元素呼应的同时，引起人们的兴致，也给购物的经历带来乐趣。

[3] 以光塑形

隔板之内是暗处，在一味的暗色调里，斜洒下的半道光强调了服装的立体感。

THE PERCEPTION OF FUTURE

感知未来

Brand Culture

|DEBRAND|

Debrand is a fashion brand owned by a Taiwanese singer, Kenji Wu. It attracts the young people in their 20s to 30s through creating the unique and new clothing. They are eager to show off their characteristics through the unique clothing.

DEBRAND是由台湾歌手吴克群创立的潮流品牌，通过创造独特新颖的服装来吸引一群20~30岁之间的年轻人，他们渴望透过独特的服装以彰显自己的个性特点。

项目信息

PROJECT INFORMATION

摄 影 师 I Figure x Lee Kuo-Min Studio

项目面积 I 200 ㎡

项目地点 I 台湾台北

设 计 师 I Michelle Wei

设计公司 I MW Design

项目名称 I Debrand

扫码查看电子书

DEBRAND
you're the one

DEBRAND
you're the one

DEBRAND
you're the one
空间创意
Space Originality

室内陈设
Indoor Furnishings

一层
The First Floor

Located in the most popular sector in Taipei, Taiwan—Ximending, and this is a fashion store of science-fiction that makes people hard to ignore. The shop windows are lined with two three-metre-high mechanical circular dormant tanks, which the strong and dazzling vision attracts the curiosity of the passersby.

With the brand LOGO as the entrance image, the hexagonal metal arch is arranged in a "人" shape with the color-changing tubes. When entering the store, the lights of the archway are activated as a result of the induction, and the linked mechanical sounds signal the start of a science fiction journey.

坐落在台北最热闹的地段——西门町，这是一间难以让人忽视的具有科幻性的服饰店。玻璃橱窗里排列着两座3m高的机械感圆柱体休眠舱，强烈而炫目的视觉效果吸引着过路人的好奇。

以品牌LOGO作为入口的意象，六角形的金属拱门与一支支变换色彩的灯管律动地排列成“人”形，当进入店内，拱门的灯光因感应而启动，连动的机械音效宣告着一趟科幻之旅即将开始。

[1] 视觉冲击

右侧是一整面沿着楼梯而上的光墙，墙上展示的武器给人十足的视觉冲击力，想象着拿起这些产品就像拿着武器登上飞船一般。“基地”门口摆放着两座圆柱体“休眠舱”，像是收藏了钢铁侠的秘密装备，让到此的客人有一种即将变身超级英雄的奇妙感觉。

灯光设计
Lighting Design

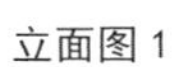

立面图 1

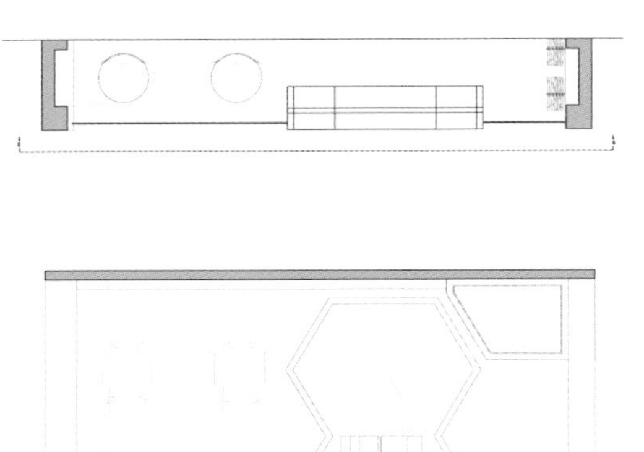

立面图 2

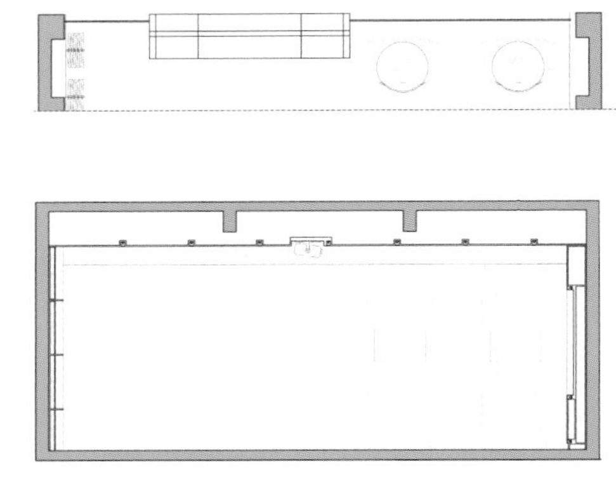

立面图 3

1 | 2
3

[1] 空间氛围

黑色的底色中，交错的白色格状光带从天花板转折到墙面，将天花与墙壁连接在一起，创造出一种神秘基地的空间氛围。

[2] 灯光装置

裸露的白色、蓝色光管既整齐又互相错落，不仅给空间带来丰富的装饰性，还增强了空间神秘的未来感。

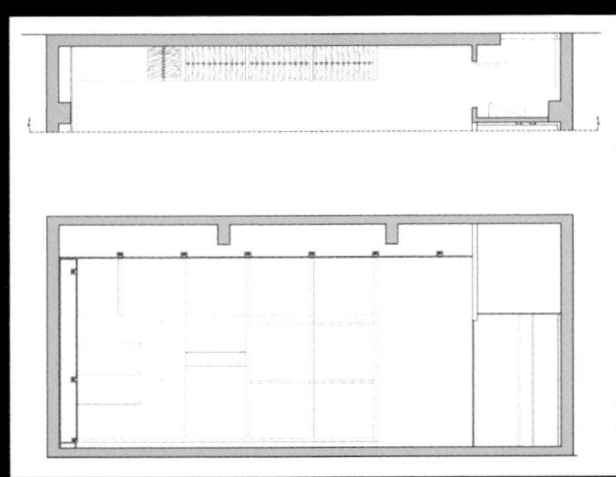

立面图 4

4
5 | 6

[3] 整体灯光效果

一楼的空间以神秘气氛为主，入口敞开的六边形门框以蓝色灯光渲染幽谧感，黑白两个“休眠舱”极具穿越感，黑色空间里天花板规格的灯带结合嵌入式小筒灯，营造宇宙浩渺的既视感。灯光效果充分契合店铺的需要。

二层
The Second Floor

Contrarily to the first floor that is covered in black, the second floor has a white theme, which the tranquil atmosphere is like a spaceship sailing in the universe. The white space uses the black lines to outline the illusion of the infinite stretching of space which can be fully immersed in the sci-fi and futuristic concept. Customers can pick the clothes from the translucent acrylic racks that look like the products are floating.

The second floor is also done a multi-functional design for the literature and art exhibitions. In the middle of the room are uneven white cubes that are used to display products and art pieces. The black lines extending from the ceiling and the wall embedded in the hooks that can hang artworks. As a whole, the shopping experience at Debrand unquestionably is upgraded and unmatched, and there is also a fantastic space for achieving your science fiction imaginations.

不同于神秘黑色的一楼，二楼是全白色的空间，宁静的氛围像一艘行驶在宇宙中的太空船。白色的空间以黑色线条勾勒出空间无限延伸的错觉，可以完全沉浸在科幻和未来的气氛中。顾客可以从透明的亚克力衣架上拿下似乎漂浮在空中的衣服。

二楼也为文艺展览做了多功能设计，在中间渐高的白色展台可以展示产品和艺术品，天花板与墙面延伸的黑色线条内埋入了可悬挂画作的轨道挂钩。整体而言，在DEBRAND不仅有无与伦比的购物体验，更是可以满足你科幻想象的奇妙空间。

室内陈设
Indoor Furnishings

[1] 镜面

二层的墙面设计了许多大面积的镜面，在线条的衬托下使得空间极具延伸性，给人明亮宽敞的视觉效果，同时也贴合了“太空舱”的主题，增强了未来感。

[2] 装饰

右侧壁柜墙面以DEBRAND品牌的东方抽象元素——甲骨文图腾装饰，增添神秘的气氛。转向左侧是以大型的店铺LOGO覆盖着的穿透更衣间。设计师使用电控玻璃作为更衣间门片，在上锁的同时，门片瞬间从透明变得雾化，并在空间中回荡着品牌理念的音效，是一个能让人感到惊喜的体验。

REFLECTION
DEBRAND
2016 SPRING
REFLECTION
DEBRAND
2016 SPRING

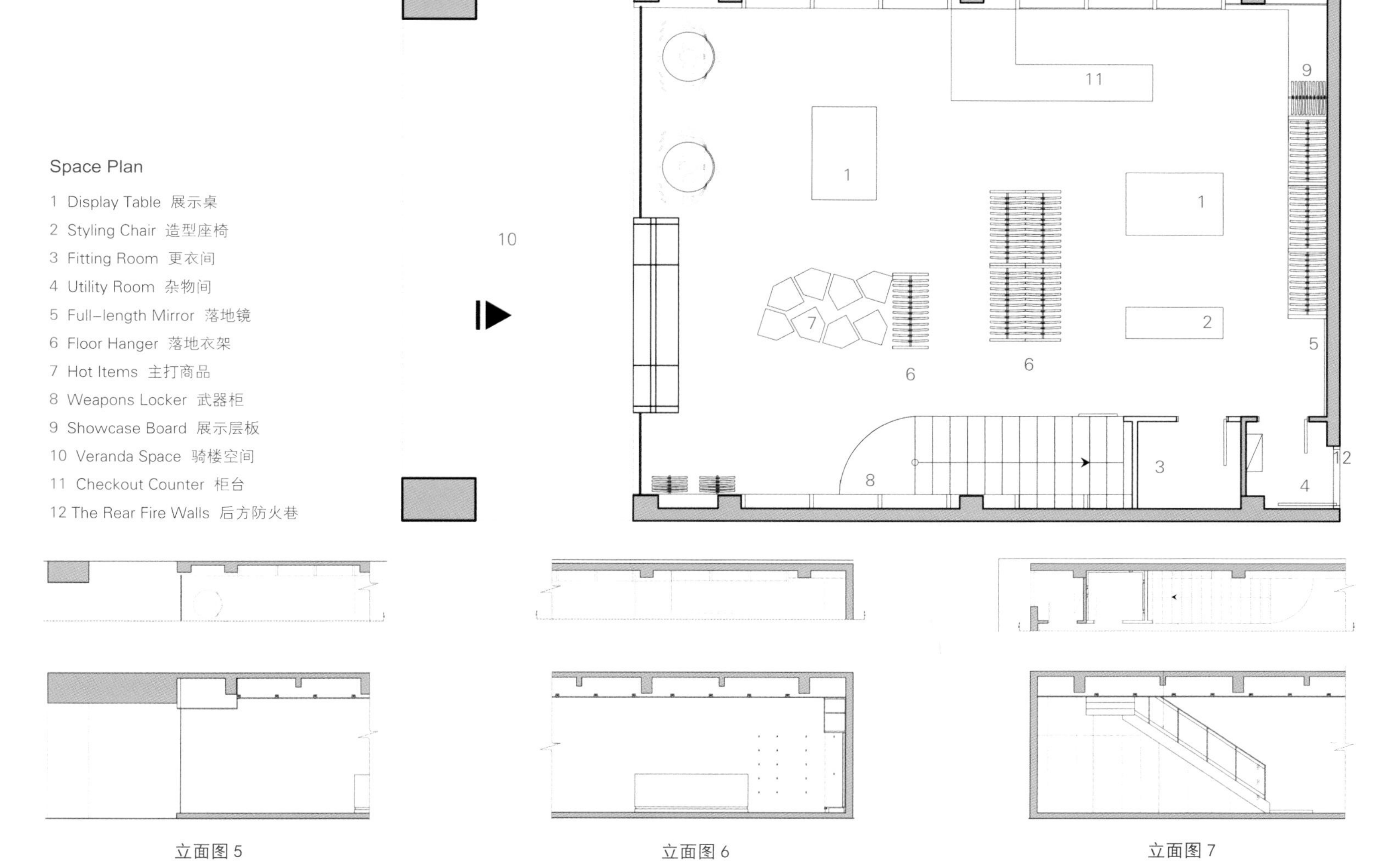
Space Plan
1 Display Table 展示桌
2 Styling Chair 造型座椅
3 Fitting Room 更衣间
4 Utility Room 杂物间
5 Full-length Mirror 落地镜
6 Floor Hanger 落地衣架
7 Hot Items 主打商品
8 Weapons Locker 武器柜
9 Showcase Board 展示层板
10 Veranda Space 骑楼空间
11 Checkout Counter 柜台
12 The Rear Fire Walls 后方防火巷
11
9
1
1
10
7
6
6
2
5
8
3
4
12

立面图 5 立面图 6 立面图 7

A FIGURED LADY·A NEW FASHIONABLE STYLE

百变女郎·穿出新时尚

Brand Culture

| JOOOS FITTING ROOM |

JOOOS Fitting Room integrates the Top 100 fashion brands in the sales list of Tmall and selects four of the most representative collections through the buyers: Mori Girl Collection, Celebrity Collection, OL Collection and Fashionable Girl Collection. The store hopes to make up the sense of emptiness of fitting brought by the modern online shopping.

就试-试衣间整合了天猫Top100销售榜上的潮牌，再由买手在其中挑选最有代表性的：森女系列、名媛系列、OL系列与潮女系列，店铺希望通过线下的体验感来弥补现代网络购物带来的试装空虚感。

项目信息
PROJECT INFORMATION

摄影师｜邵峰
项目面积｜1850㎡
项目地点｜浙江杭州
设计团队｜刘欢、任丽娇、贾媛媛
主案设计师｜李想
设计公司｜唯想建筑设计（上海）有限公司
项目名称｜就试 试衣间

扫码查看电子书

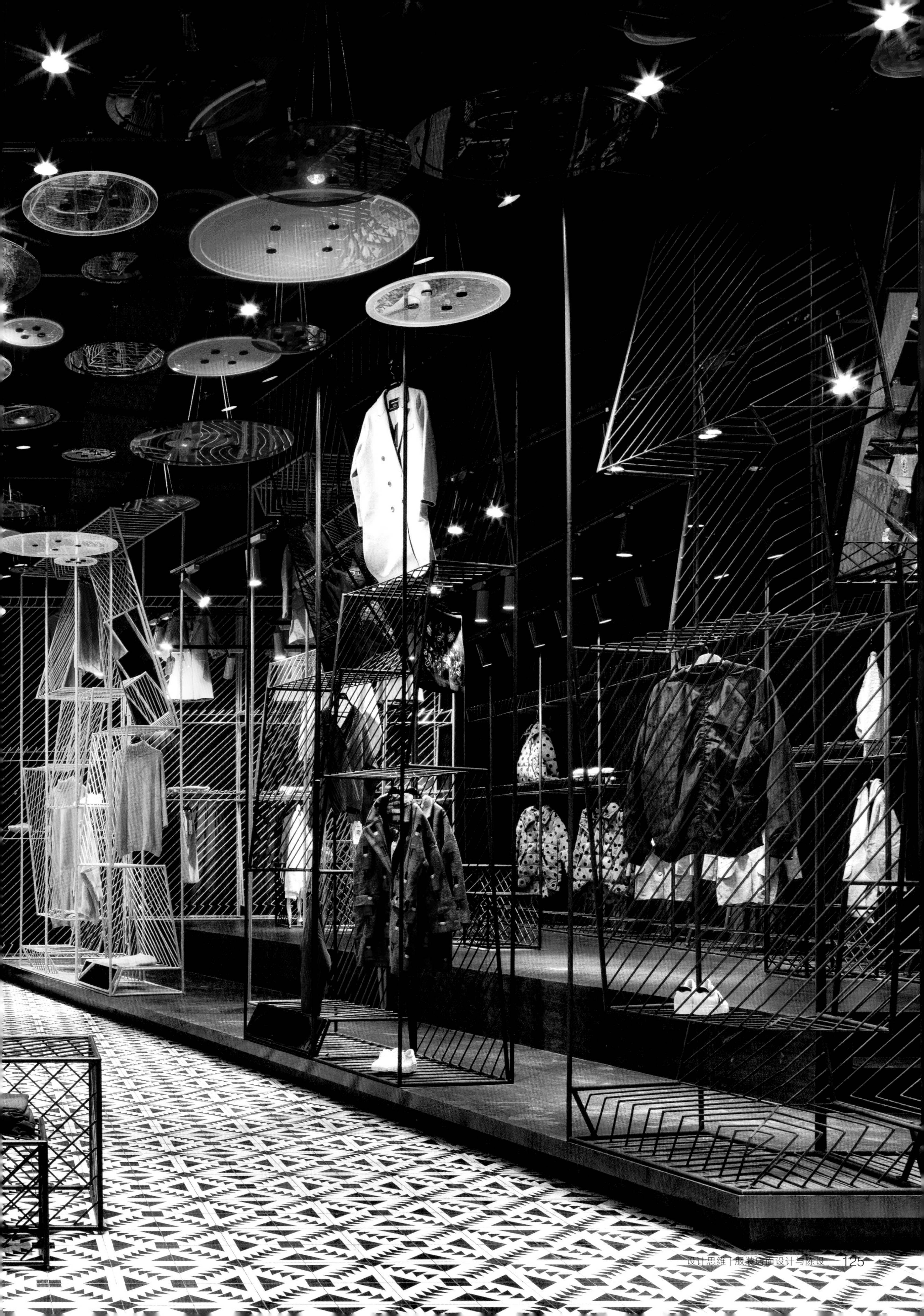

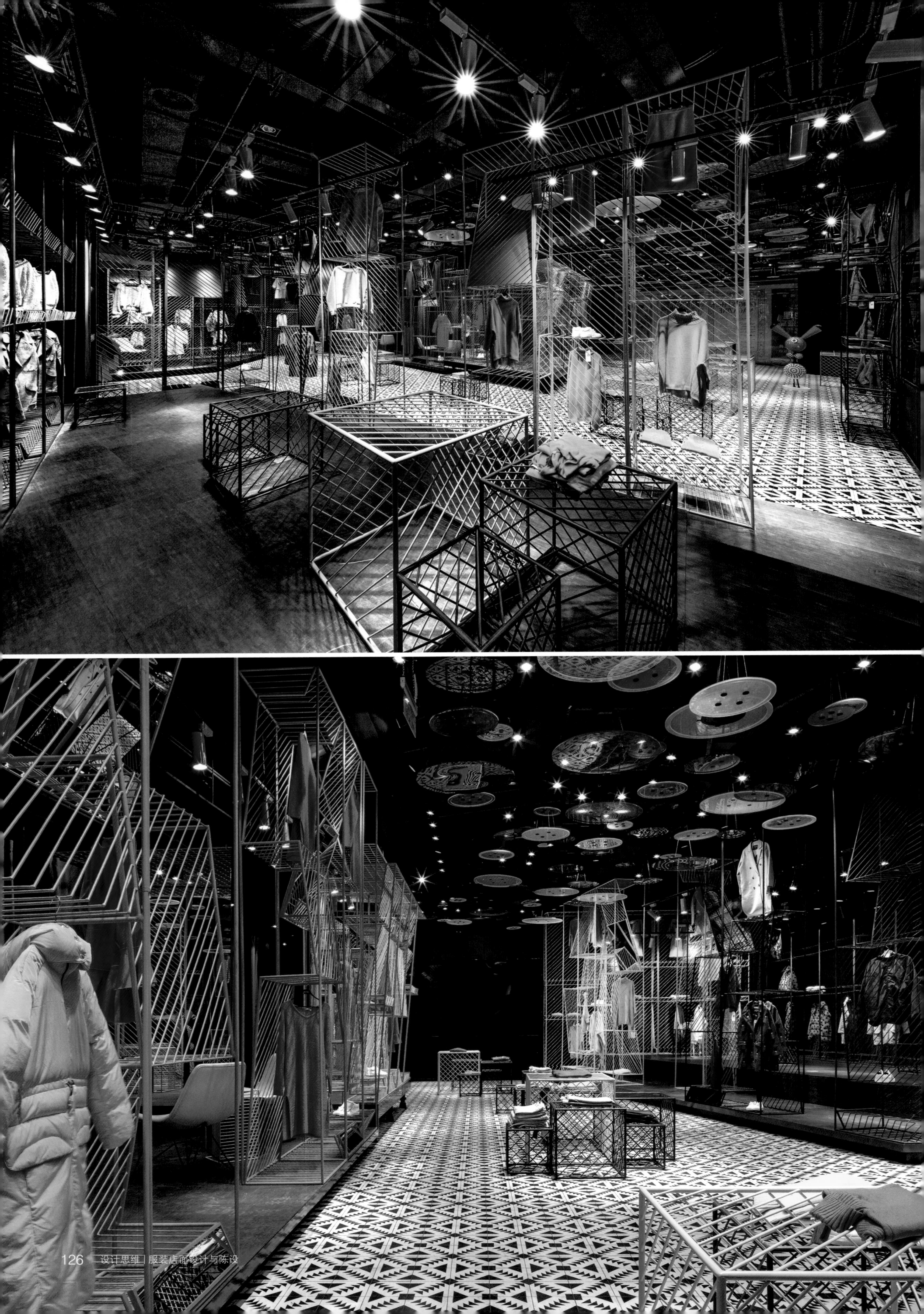

|潮女区|

Fashionable Girl Collection Area

Hangzhou JOOOS Fitting Room is located at the door of the ground floor of the Commercial Street Phase II on Xingguang Avenue. The designer integrates the social and cultural connotation of the four types of women's brand clothing. The dressing philosophy of women's clothing is interpreted by using different techniques in designing the four spaces, and a new shopping experience is created through multiple possibilities of the display.

杭州就试-试衣间，位于杭州星光大道商业街二期一楼的大门口位置。设计师整合了女装品牌四大类型服装背后的社会与文化含义，通过四个空间不同的设计手法来演绎女装背后的穿衣哲学，并且通过多种陈列的可能性创造了全新的购买体验。

室内陈设
Indoor Furnishings

1 | 2

[1] 色彩突出

以黑色空间为背景突出琳琅满目的彩色服饰，突出空间主体——服装，与此同时还增加了空间层次感。

[2] 多色彩搭配

在这强调个性的空间里有大面积撞色，设计师利用前后折度将铁架子分割成双面挂衣的结构，用彩色与折线的拼织来打造幻彩个性的空间，以此来呼应这个空间里服装所展现的独特性。

灯光设计

Lighting Design

[1] 情境营造

潮女区无主灯，光源主要来自天花板的射灯，无固定排序的射灯打在衣架上，迅速形成视觉片区，使得空间即使色彩丰富也完整而不显杂乱。

[2] 空间装饰

设计师在天花射灯下设计许多彩色的圆形大纽扣作为装饰，让色彩在空间中上下呼应，此外，通过灯光把纽扣绚烂的色彩渲染到空气里，让独立的彩色衣架糅合为空间的一个整体。

Celebrity Collection Area
名媛区

Celebrity Collection Area — many delicate golden cages are designed, and clothes racks are placed at both inside and outside of the cage. Seen from a distance, they are like princess bubble skirts being well protected. The waving arc platform adds fun to space and makes it lively and lovely. The fitting room is hidden skillfully inside the “bubble skirt” of the curved mirror surface. Meanwhile, each fitting room area provides a make area, rest area and selfie area, waiting for the visit of every princess.

名媛区里有一个个精致的金丝笼，衣架被放在金丝笼内外两侧，远看仿若被精心呵护的公主蓬蓬裙。高低起伏的弧形平台在增加趣味性的同时又不失活泼可爱。试衣间被巧妙地隐藏在弧形镜面的“蓬蓬裙”内侧。同时每个试衣间区域也提供梳妆、休憩、自拍区，等待着每一位公主。

室内陈设
Indoor Furnishings

1
2

[1] 服装展示

空间色彩与服装色调需要默契搭配，柔和的偏莫兰迪色系的服饰摆在浅光里，没有弱化服装的地位，反而增强了服装的柔美气质，符合“名媛”对高雅品质的追求。

[2] 动线

由“金丝笼”的立面构成了蜿蜒的空间动线，在带轨道多头射灯一圈圈配合下增强了“金丝笼”造型的效果。即使没有购买计划，在香槟色的空间氛围中走马观花、停留拍照也是一番美事。

Get through the large arch in the main entrance, and you will be in Mori Girl Area. The milky white textured walls and the white floor create clean and white space. Bamboo poles are set up, and the hemp rope is used to connect them, clothes racks are thus formed. The mirror is hidden in the facade of an angle formed by two bamboo poles.

通过主入口区大大的门拱进入到森女区。乳白色肌理墙面、白色地面、一个洁白干净的空间，支立着一根根竹竿并以麻绳做连接，而衣架也就此成型。穿衣镜则隐藏在两根竹竿的夹角立面上。

|森女区|
Mori Girl Area

|OL 区|
OL Collection Area

OL Collection Area—dark gray floor, concrete art paint walls and frame track light make the whole space concise and reserved. Adding of fireplace and wood veneer has softened the texture of space. While being decorative molding and multifunctional clothes hanger at the same time, every clothes rack is a perfect combination of form and function.

OL区采用深灰色地板、混凝土艺术漆墙面，框架轨道灯让整个空间简练、稳重，作为点缀的壁炉、木饰面柔和了空间质感。每个衣架即是装饰性线条又是多功能衣架，形式美与功能性完美结合。

空间创意
Space Originality

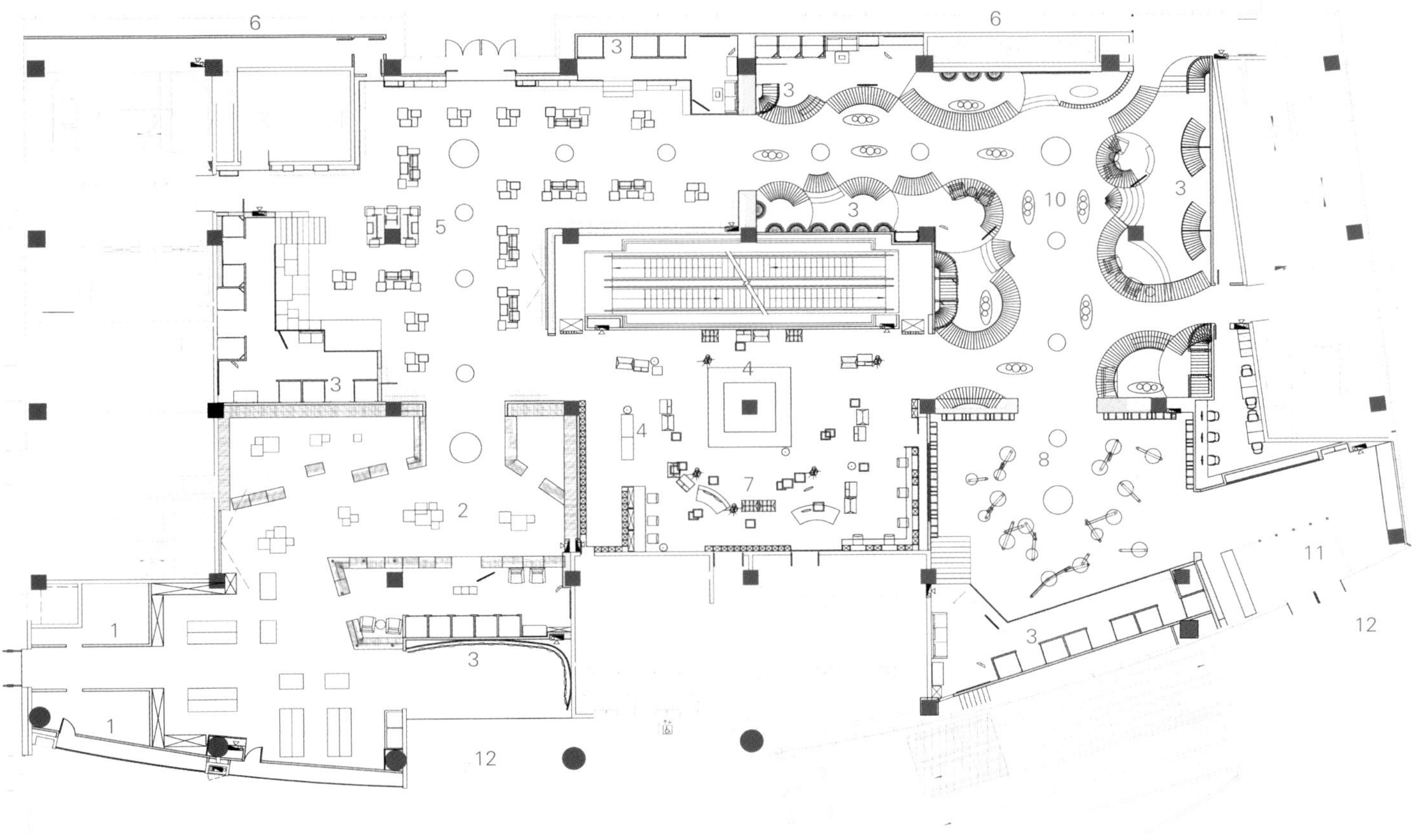

Space Plan

1 Storeroom 储藏室
2 Tide Female Area 潮女区
3 Fitting Room 试衣间
4 Cashier 收银台
5 OL Area OL 区
6 Showcase 展示区
7 DIY Area 自定义区
8 Mori Girl Area 森女区
9 Cashier 收银台
10 Sociallite Area 名媛区
11 Service 服务区
12 Entrance 入口

室内陈设 Indoor Furnishings

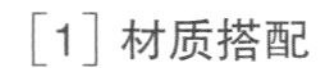

[1] 材质搭配

为了营造职业干练的空间气质，设计团队把OL区做成一个黑色空间，加入黑色的金属线条描摹空间线条，让展示区轮廓分明。其次，使用浅木色展柜、桌椅来温暖灰色空间的冷落枯燥。

[2] 有趣的几何

大部分职场服装空间做得非常规整，但在这里，我们除了看见职场女性的干练之外，还有温柔、时尚、稳重且魅力十足。灰色的展柜以黑色描边，采用岛式陈列，搭配木色立体几何创造出一个不规则整体，体现职场女性对“不走寻常路”的个性化追求。

灯光设计
Lighting Design

[1] 无主灯

射灯是典型的无主灯、无规定模式的现代照明灯具。天花板的轨道窄光束射灯在这里起到主导作用，让空间稳重又简练。

[2] 横向空间层次

展示区通过射灯聚光对服装进行点题。设计师在嵌入式展示区的上部做向外凸出的金属衣架，内嵌部分使用暖光加大视觉深度，以此增强空间的层次感。

[3] 空间氛围

整体空间是浅灰色，灯光如果太强会破坏空间浅浅的稳重感。墙面上略带工业风的金属框架除了做衣架外，还起到了装饰作用，活跃空间平静缓和的节奏。

[4] 纵向空间层次

OL区做了下沉式展示区，搭配平地上凸起的展柜，丰富了空间节奏，同时自下而上拉伸了视觉效果。

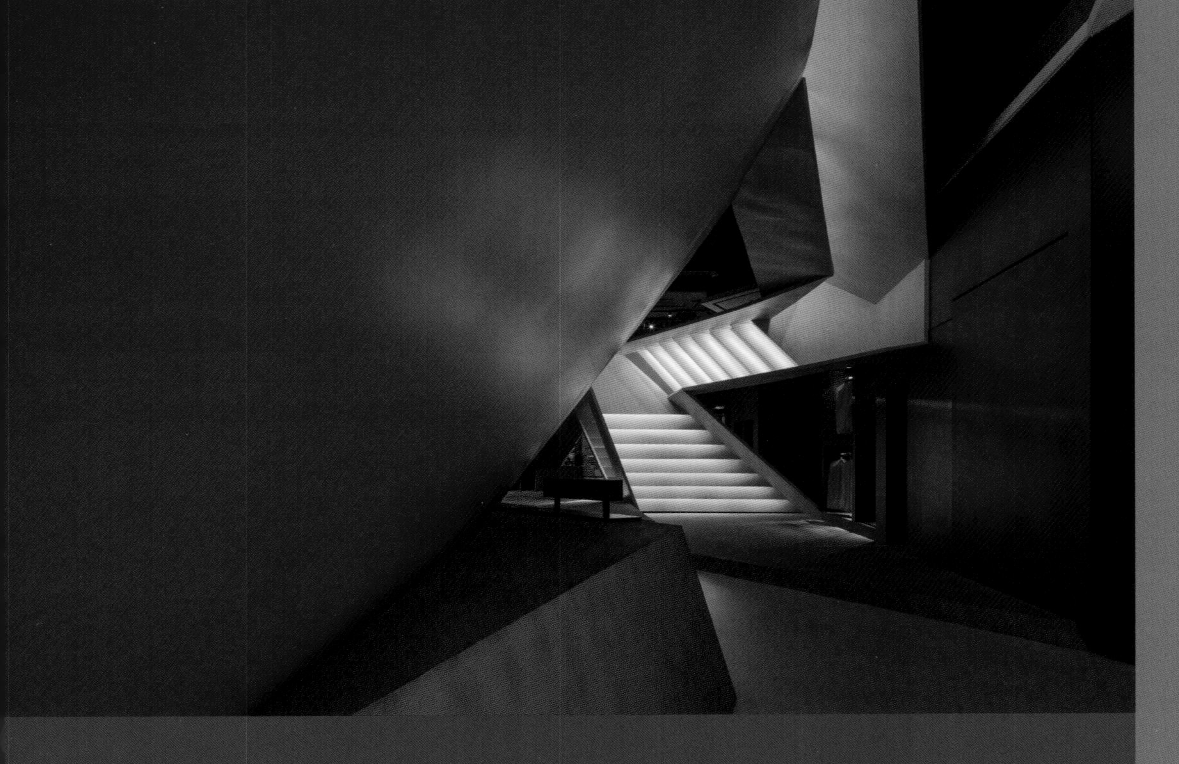

DISPLAYING YOUTH AND PERSONALITY

张扬青春个性

| ROARINGWILD |

ROARINGWILD confronts life by the "Roar" attitude, especially the young people who dare to pursue their dreams, not to live for a living and not to survive for the material things. ROARINGWILD is not only designing but also being a pioneer. All ideas are conveyed through the clothing as a carrier, constantly enriching the brand's content and creating a life attitude that belongs to ROARINGWILD.

ROARINGWILD通过咆哮"Roar"的态度来面对生活，尤其是年轻人能够敢于追逐梦想，不是为谋生而生活，也不是为物质而生存。ROARINGWILD不仅仅是做设计，更多的是做一个先行者，所有的想法都通过服装作为载体来表达传述，不断丰富品牌内涵，创造属于ROARINGWILD的生活态度。

项目信息
PROJECT INFORMATION

项目地点 | 广东深圳
施工单位 | 深圳市尚治装饰工程有限公司
协作设计 | Carving Tang 唐嘉颖等
主案设计 | Kingson Liang 梁永钊
UNIWALK | ROARINGWILD | 方城店
项目名称 | ROARINGWILD

扫码查看电子书

ROARINGWILD®
ROARINGWILD®
ROARINGWILD®
ROARINGWILD®

ROARINGWIL
ROARINGWI

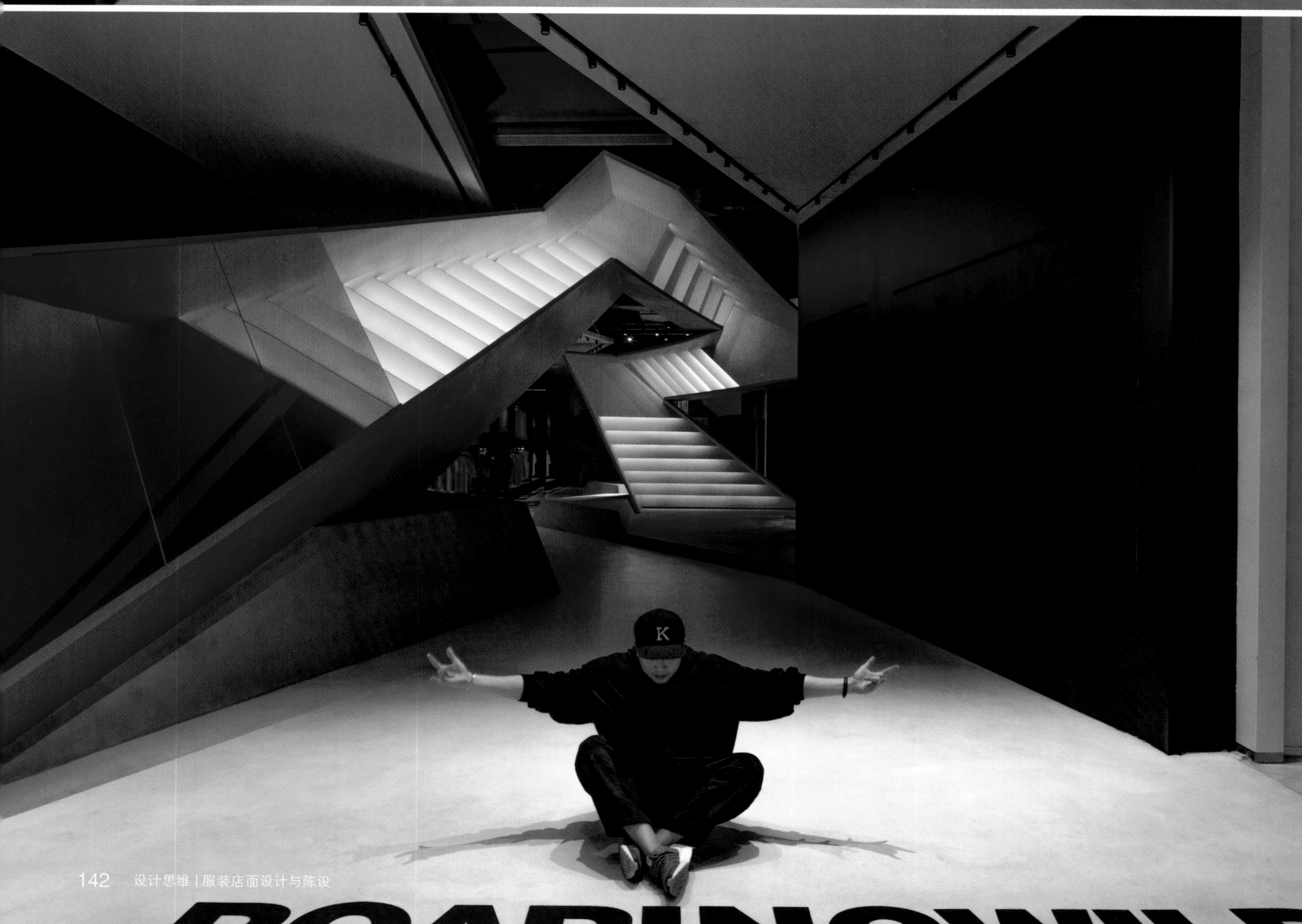

空间创意

Space Originality

An all-red giant device runs through the whole space and becomes the existence of vision and art. From a front perspective, it also forms the R shape of the first letter of the brand name. The device uses the red steel plate as the main material, only the area near the entrance using the gloss red glass material.

The geometric device is full of a futuristic sense. And the interior fills with the contrast color tone of red and black: red symbolizes passion and eruption, while black is calm and restrained. The perceptual blood grows in the rational enclosure, and these two colors are balanced in the confrontation.

一座通体透红的巨型装置贯穿整个空间，成为视觉与艺术的存在，从正面的角度，亦形成品牌名的首字母“R”。装置以红色钢板为主材，只在靠近入口处的部分采用光泽的红色玻璃材质。

几何装置充满了未来感，室内充斥着红与黑的撞色基调：红色是激情和爆发，黑色是沉稳和内敛。感性的血液在理性的围合中滋长，两种颜色也在对抗中达至平衡。

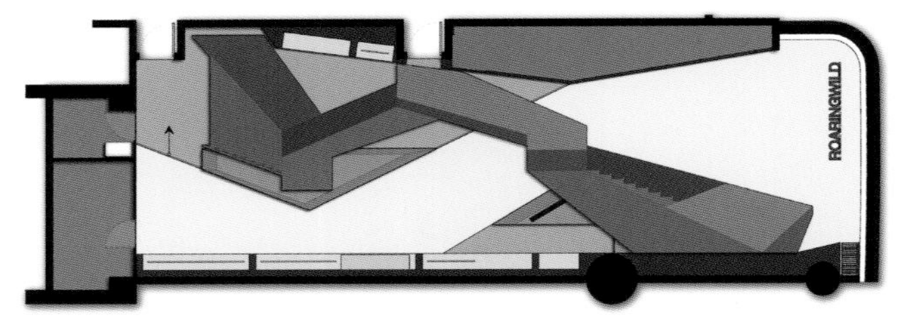

Space Plan

ROARINGWILD

室内陈设

Indoor Furnishings

[1] 灵活的空间开合

入口通道右侧的黑色烤漆板墙面，闭合时可用作商品陈列，或当作投影墙，开启时可扩大空间，拓展更多内容。灰色水泥材质收银台位于红色装置下方一角，在红黑主色的氛围里犹显安静。

[2] 制造视觉焦点

设计师以凌厉的笔法构建空间的感官错觉和视觉冲击，黑色的背景墙十分突显服饰魅力。

[1] **打破传统零售的枯燥程式**

空间中的“R”形装置以楼梯的形式呈现，每一步楼梯都安装了灯带，使得整个装置立体而壮观，并制造出张扬不羁的空间印象。

[2] **视觉平衡**

空间以黑色为背景反衬服饰的色彩，零星的射灯在天花上起到点缀作用，使得神秘的黑色空间显现出时尚与浪漫的气氛，与此同时，与“R”形的大型装置形成点面结合的视觉效果。

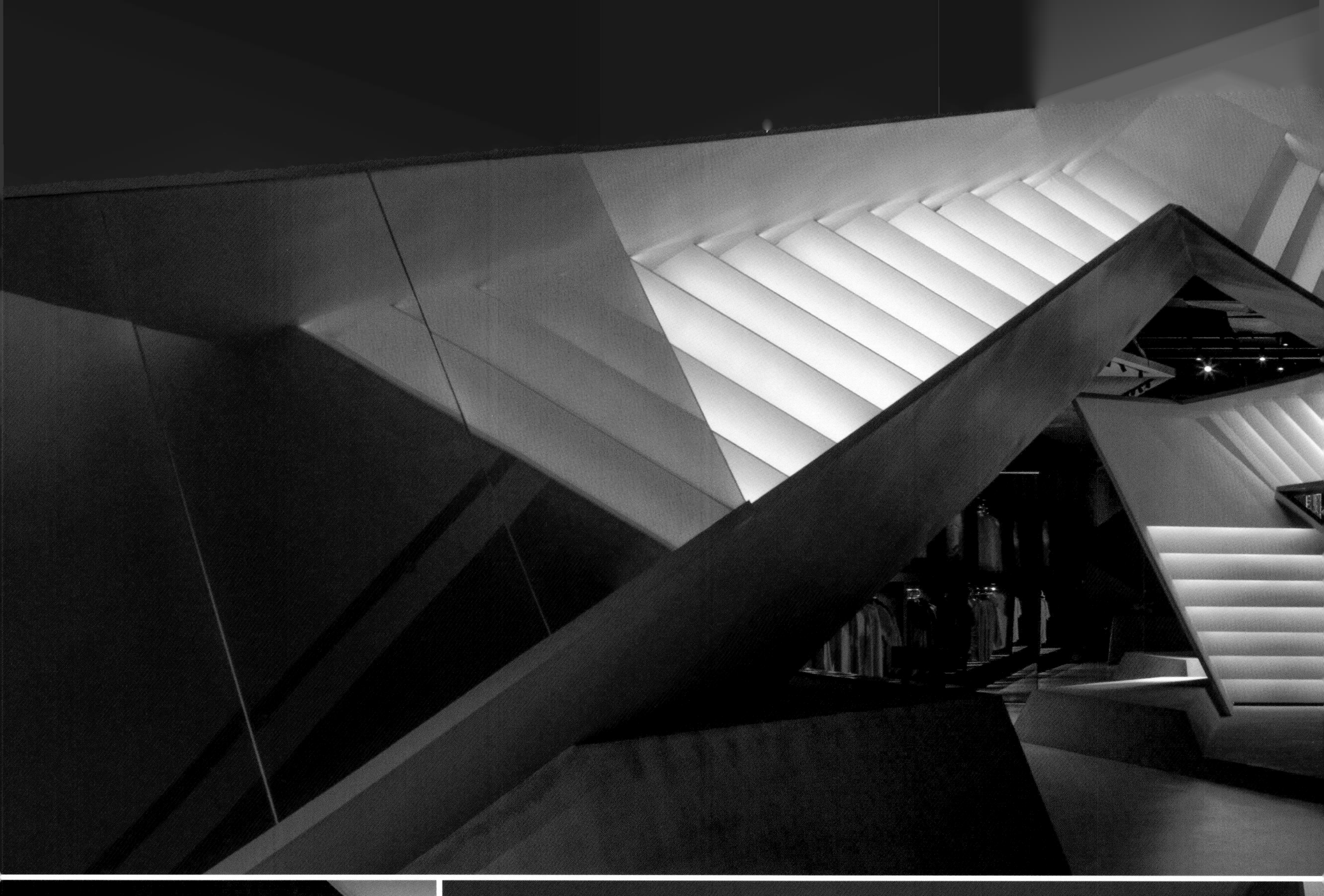

［3］气氛营造

纵使“R”装置给人惊喜的体验，但加入光效的阶梯给人更清晰的层次感，让人在不知不觉中受到引导，从而深入到店铺内部。

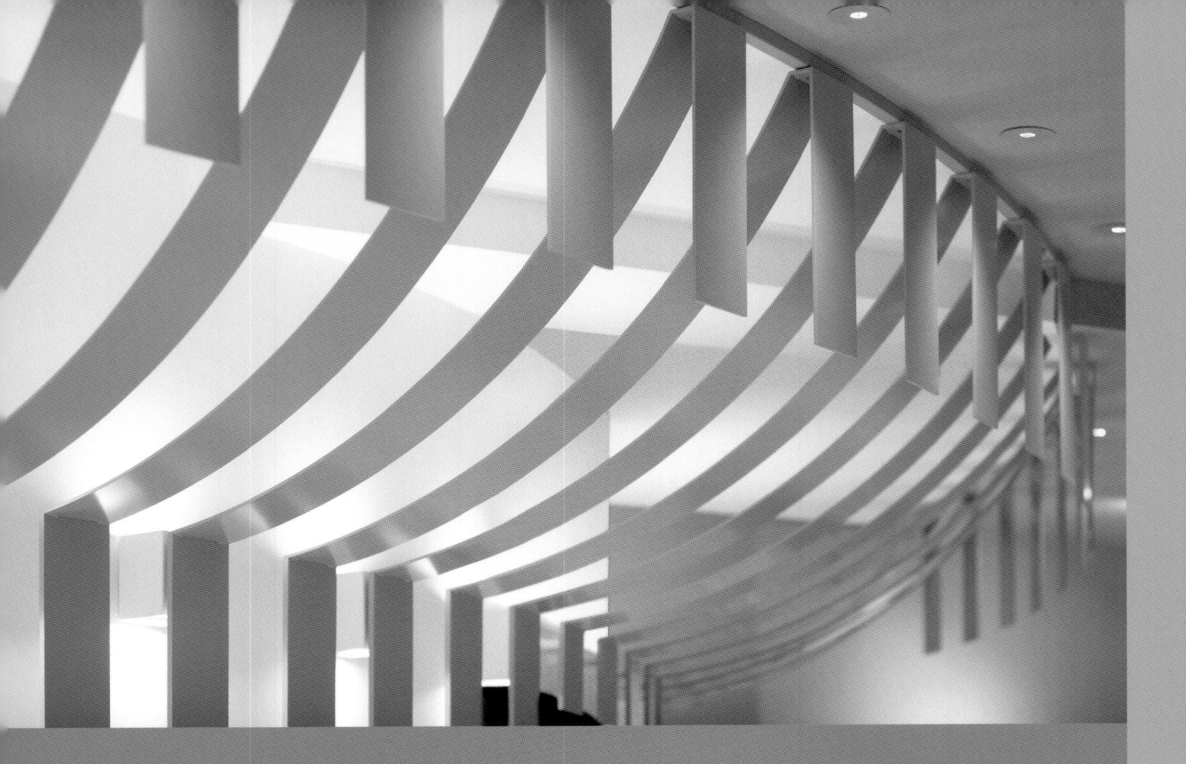

THE WOMEN'S FASHION SHOP WITH A MOBILE AMUSEMENT PARK

装着流动游乐园的女装店

Brand Culture

| KEY TO STYLE |

KEY to STYLE is the women's fashion floor within the main building of the Seibu Shibuya department store. It has a casual and relaxed atmosphere for young female customers, with a sense of leisure for young customers as a feature.

KEY to STYLE位于日本西武百货主楼内的时尚女装楼层，它以随意、轻松的氛围面向年轻的女性客户群体，以针对年轻客户的休闲感为特色。

扫码查看电子书

项目信息
PROJECT INFORMATION

Project Name I KEY to STYLE

Design Company I nendo INC.

Designer I Oki Sato

Project Location I Tokyo, Japan

Area I 1,047 ㎡

Photographers I Interior pictures by Takumi Ota, fixture pictures by Akihiro Yoshida

ANREALAGE
ANREALAGE

ANREALAGE
ANREALAGE

空间创意
Space Originality

The designer wanted to create a dynamic setting which can echo with the "cozy European park" style of the "COMPOLUX" area on the third floor of the annex building that we designed for the Seibu Shibuya department store in 2013. Hence, the designer decided on the "mobile amusement park" that is held in parks as the design theme.

设计师希望创造一个连动设置，使得KEY to STYLE能与设计公司2013年为西武百货附属楼三层的“COMPOLUX”区的“欧洲闲适公园”主题风格相呼应。因此，设计师决定将公园里的“流动游乐园”作为KEY to STYLE的设计主题。

KEY to STYLE

three dots
ANREALAGE

室内陈设
Indoor Furnishings

[1] 综合陈列

空间以天花板的蓝色为基调，再辅以一些亮眼的设计元素，同时在个别区域使用木质构件来强化产品的色彩和质感。在KEY to STYLE女装区的地板设计中，设计师加强了色彩的对比，并将地砖的纹理同条形的图案结合起来。

销售区的设计灵感来自于马戏团的帐篷，品牌位置的排布很像市场上的一个个摊位，配饰区域的设计则是受到了传统马车式样的启发。

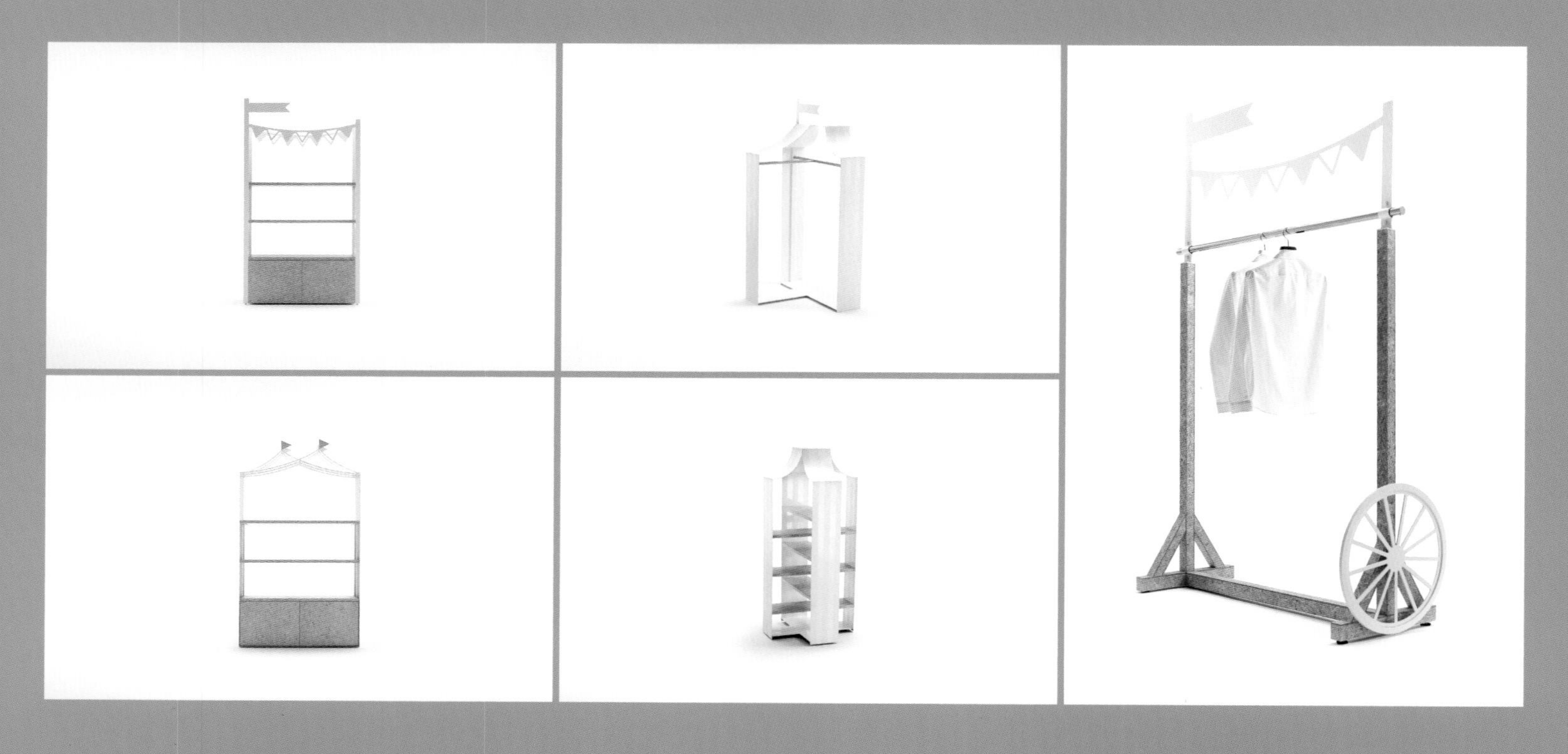

LE CIEL BLEU

灯光设计
Lighting Design

[1] 灯光辅助布局

销售区位于自动扶梯的四周，顾客可以环绕一圈，不至于重复路线或迷路。镜面灯的设计给空间增加了游乐园般的趣味性。同时，设计师还设计了两条对角线走道，提高人流的流畅度。

[2] 明暗设置

动线上的光效相对较暗，在展柜、整体陈列区的灯光则更明亮。白色的灯光在镜面的配合下给人清新通透的视觉效果。

THE ELEGANT PINK DANCING WITH THE UNRULY GEOMETRIC ELEMENTS

高雅粉与几何元素的不羁共舞

|Novelty|

Novelty is a global woman's fashion clothing brand founded in 2013 by two graphic designers. Its main customers are the fashionable and young women in the modern metropolis, and it mainly produces well-selected and individualized items that represent the vanguard of the trend. Renowned for its iconic fashion sense, Novelty has expanded to a full collection of ready-to-wear and accessories including shoes, handbags, small leather goods, scarves and fine jewelry, and has become one of the most important apparel stores in the region.

Novelty是一个全球性的女性时尚服饰品牌，由两位年轻的平面设计师创立于2013年。其以现代都市里时尚年轻的女性为主要客户群体，主营精心挑选的、个性的代表潮流先锋的物品。凭借着标志性的时尚感，Novelty的经营范围已经从成衣和饰品扩展到鞋、手袋、小件皮革制品、丝巾和高级珠宝等，并成为所在地区最重要的服装商店之一。

扫码查看电子书

项目信息
PROJECT INFORMATION

Project Name | NOVELTY

Design Company | ANAGRAMA

Designer | ANAGRAMA

Project Location | Mexico

Main Materials | Marble, metal, etc.

Photographer | ESTUDIO TAMPIQUITO

NOVELTY

空间创意
Space Originality

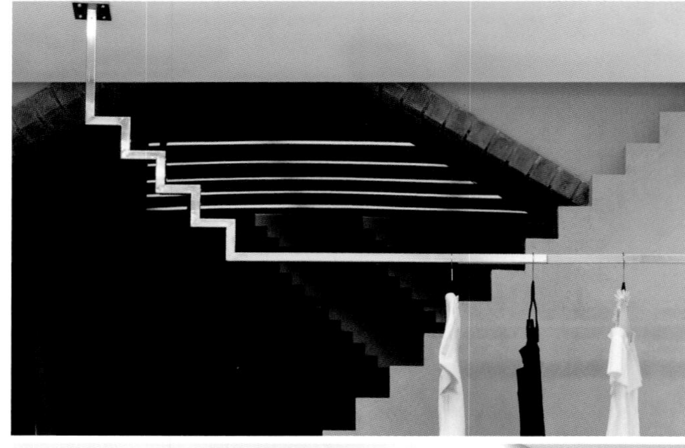

It was the second interior design we made for the Novelty. Designers got inspired by the works of Mexican architect Luis Barragan and Spanish architect Ricardo Boffil and tried to merge those two different styles into one concept. The store uses pink as the main color, and some saturated colors as complementary colors, interspersed with various geometric shapes and matched with the ladder-like design, all of these constitute a new style of brand design, present a vivid, gorgeous new thinking and show the passion, nature and vitality. The spacious, bright visual effects and the fluency of the physical space together create a modern feminine space, providing an exciting experience for visitors wandering around the store.

这是设计师第二次为Novelty品牌服务，其从墨西哥建筑师路易斯・巴拉甘和西班牙建筑师里卡多・波菲尔的作品中获得灵感，尝试将二者的不同风格合并成一个概念。店铺以粉色为主色调，用一些饱和色互补，各种不同的几何造型穿插其间，搭配阶梯状的设计，构成了品牌设计的一种新风格，呈现出活泼、华丽的新思维，表现热情、率性和活力。宽敞、明丽的视觉效果以及物理空间的流畅性共同造就了一个现代的女性化空间，为徜徉于店铺内的访客提供一个非常有趣的体验。

室内陈设
Indoor Furnishings

[1] 透过橱窗，树立良好的第一印象

在临街的玻璃窗前，两位时尚年轻的模特穿着不同款式的黑色裙子，以不同的舒适姿势站着，眼神都若有所思，上方是品牌标志，背后是精品店具有代表性的一角，雅致、有品位，调动潜在消费者的视觉神经。

[2] 丰富的展示方式，让顾客有愉悦感

在背景墙上，设计师用不同的结构作为装饰，简单中透露着利落与清爽。在此基础上，商品展示完全因结构而发生变化，或是置于四周阶梯，或是挂在阶梯状陈列架，或是放在中央的陈列台上，既有对称的陈列，也有不规则的摆放，给予空间轻松愉快的气韵。

[1] 以照明效果突显空间特色

几何形状无疑是这个空间最出色的特点之一，设计师通过灯光的对比和阴影的表现，强调几何形状。

[2] 光的形状

光除却本身造型之外，因为建筑结构的关系，也能幻化出有意思的形状。而这些形状与结构组合在一起，能产生神奇的效果。

灯光设计

Lighting Design

[3] 发挥光线作用，让购物变为享受

室内照明以平行嵌入式的长条状灯具为主，给粉色的商店亮丽之感，对服装而言也具有良好的显色性。在弧形、阶梯状等结构的交界处，光线稍有聚拢，增强空间的层次感。在灯光的作用下，大理石、金属等各种材料表现出店铺线性流畅的物理特征，饰面的坚固属性也赋予空间暖意和女性化的柔和，消费者购物的时光也变为享受的过程。

THE PLANET OF A TENDER IDEALIST

一个温柔理想主义者的星球

品牌文化 Brand Culture

MILANO BOTIQUE

Milano Botique is a buyer shop created by Nic, which prefers the European-American style. Nic takes Milano Botique as a basis point for a new departure from the part. Therefore, different from the previous store form, the aim of Milano Botique is becoming an innovative future store. Nic hopes that people who come here can feel that Milano Botique is close to their ideal store but also hopes to establish a link between goods, design and people through this store.

Milano Botique是由Nic创建的一家偏欧美风格的买手店。Nic将Milano Botique视为一个基点，一个告别过去重新启程的基点。所以，区别于以往的商店形式，成为一个革新性的未来商店是Milano Botique贯穿始末的宗旨。Nic希望来到这里的人们可以感受到Milano Botique是贴近他们理想的商店，更希望通过这家店铺建立起商品、设计与人之间的联系。

项目信息

PROJECT INFORMATION

项目名称 | Milano Botique买手店

设计师 | 夏慕蓉、李智

项目地点 | 上海

摄影师 | Unitu、刘洋

扫码查看电子书

Milano Botique

Short Be Cool

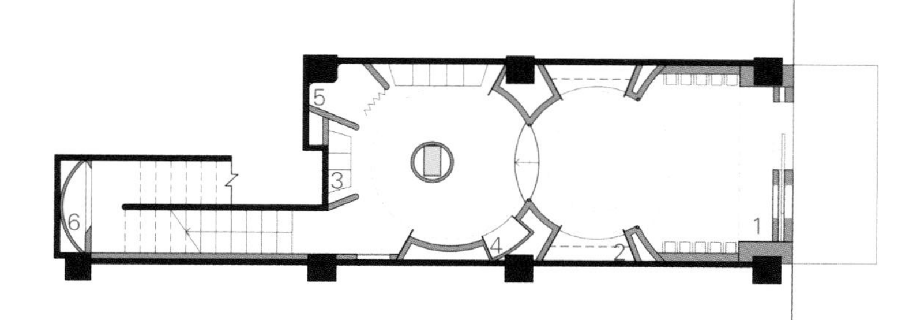

1st FLOOR

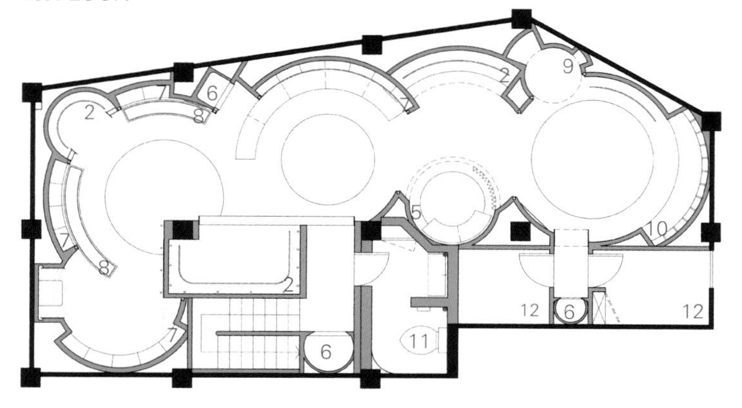

2nd FLOOR

1 Glasses Area 眼镜区
2 Clothing Area 成衣区
3 Shoes and Hats Area 鞋帽区
4 Checkout Counter 收银台
5 Fitting Room 更衣间
6 Niche 壁龛
7 Bag Area 包包区
8 Jewelry Area 珠宝区
9 Treasure Room 珍宝室
10 Lounge 休息区
11 Toilet 卫生间
12 Storage 储藏室

I need a
huge amount of mo

空间创意
Space Originality

NIC is a cool rapper with T-shirt and tattoo. However, there is a tender idealist inside. It will be difficult to be understood by the public at the beginning. Nic says goodbye to the past with his new boutique shop.

It is his planet.

The first floor consists of three elliptical spheres. The luminous ceiling provides the interior of the soft light, depicting the sharp outline of the geometry. The second floor is based on a series of arcs. The ups and downs of the plane delineate the layers of space, corresponding to the different display areas.

酷酷的Nic像个rapper，白色T恤和花臂，内心却是一个温柔的理想主义者。每个新事物的开始都会有难以被大众理解的困境，新启程的买手店，业主Nic想要告别过去。

这是他的星球。

一层以三个椭圆的球为形，向上收窄。发光顶棚提供室内柔和的光线，刻画出几何体鲜明的轮廓。二层以一系列圆弧为基准面，平面上起伏的轮廓划分出空间的层进，对应不同商品的展示区域。

室内陈设
Indoor Furnishings

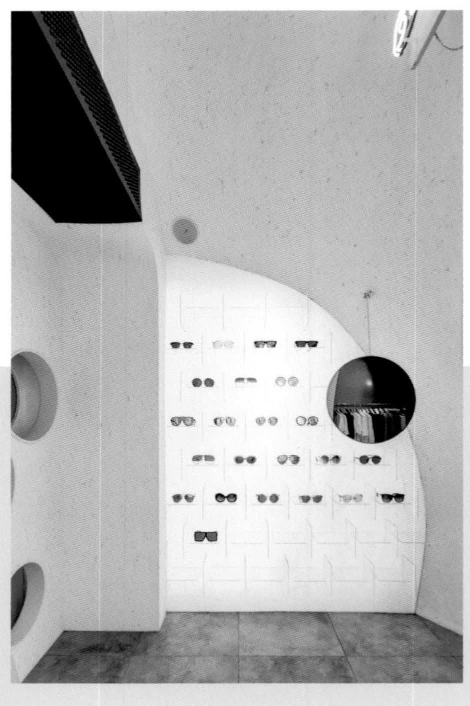

1 | 2
3

[1] 母题——圆

圆形是延展的又是向心的，是完整的又足够灵活。圆形组合兼容场地曲折的边界，隐藏复杂的结构，并提供了一个强大的主观几何秩序。圆弧在垂直方向发展为铺地、坐凳、展架、墙体、屋顶和灯具的一系列演绎，赋予空间一种清晰明辨的特征。

[2] 几何秩序

圆，一种带有超越物性隐喻的符号。因为水磨石、不锈钢和粉色软包的材料组合，它在这里没有表现出纪念性的特征，反而呈现出一种灵动诗意的柔软感。几何秩序是设计师切入每个复杂现状的有力工具，自然叙事是他们当下关注的空间理论模型。

[1] 时光隧道

廊道以柔和的白光配合白色空间，经过窄长的白色楼道，仿佛穿越宇宙的时光隧道，从一个星球步入另一个星球。

[2] 明亮的空间

整个店铺以白光为主，给人干净敞亮的临场感受。每一个展示区都有自己的背景色，在白色的整体空间里突显出展示品的主体地位。天花做不规则弧形设计，以灯带配合射灯的组合，给人向上的视觉延伸感。

[3] 圆弧主题

空间中“圆”的主题通过灯带来突显，大大小小的圆形经过灯光的修饰变得更加立体。隐藏的灯带使得可见光效而非光源，给空间以柔和的灯光氛围。

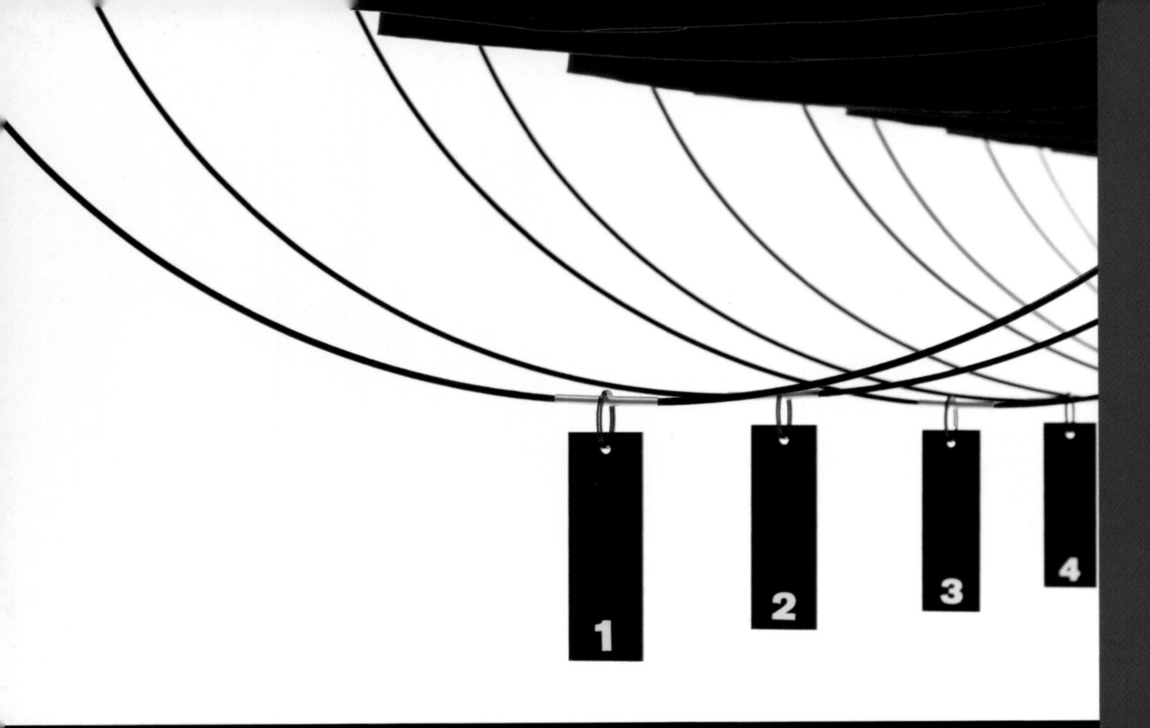

WANDERING IN THE BRIGHT SHADOW

游走于明净与光影照拂之中

Brand Culture

LOREAK MENDIAN

Loreak Mendian is a Spanish city brand, born in the Basque region, which is the "founder of leisure fashion". Loreak Mendian started in 1995 as a small shop in the port of San Sebastian, where Xabi Zirikiain and Victor Serna made and sold T-shirts themselves. Now, their designs are present in such places as France, Switzerland, Australia or Japan, and they also have the new brand image. The brand is dedicated to drawing inspiration from everyday to create a street style, the exquisite tailoring of T-shirts to give customers a comfortable and natural enjoyment. At the same time, the clothes are strongly influenced by the Spanish style. Designers like to reinterpret elements from Basque culture and interpret the "modernist basque design" as the thought.

扫码查看电子书

项目信息 PROJECT INFORMATION

Project Name | FLAGSHIP STORE LOREAK CHICA

Design Company | Pensando en blanco

Designer | Aurora Polo

Lighting Design | Islada lightning

Window Display | Jon Ander Beloki

Area | 270 m²

Interior Design Photographer | Iker Basterretxea

Shop Window Photographer | Pablo Axpe

Ura Jacket
LOREAK
Ura
Ura, agua en Euskera, es el nombre de esta chaqueta que tienes delante de ti,
diseñada para protegerte y no empaparse en días lluviosos.
Sabemos mucho acerca de la lluvia por aquí. Estamos bien familiarizados
con el sonido de las gotas de la lluvia cayendo sobre nuestras chaquetas
Obsesionados por encontrar el sonido único de la lluvia...
Drip, drip, drip...
Ura, water in the Basque language, is the name of this jacket in front of you,
and which has been designed to protect you and not get soaked in rainy
days. We know quite a lot about rain in here.
We are fairly familiar with the sound of the raindrops falling on our jackets.
We are obsessed with finding the unique sound of the rain...
Drip, drip, drip...
FW17
No.03
loreakmendian.com

Loreak Mendian是一个西班牙城市品牌，诞生于巴斯克地区，是“休闲时尚缔造者”。其于1995年开始在圣塞瓦斯蒂安港开设小商店，由缔造者Victor Serna和Xabi Zirikiain自己制作并出售T恤，如今品牌设计在法国、瑞士、澳大利亚、日本等地均有出现，也有新的品牌形象。该品牌致力于从日常汲取灵感来缔造街头风格，T恤精致的剪裁带给顾客舒适、自然的享受，同时服饰受到西班牙风情的浓重熏陶，喜欢用巴斯克文化重新诠释元素，诠释缔造者所想的“现代主义巴斯克设计”。

This shop has a dynamic appearance, with the surprise factor taking on a prominent role. The woman's shop is a space with two floors. This environment is home to vertical lines, where the perimeter lighting delimits the white walls' volume, treated with a unique texture. In this simple space, there are different areas remarkable in the shop. For example, on the ground floor, there is an area delimited by 1.8 meters tall blue glass, providing the special treatment for collections. The blue transparencies bring freshness and plasticity, combining it with the different heights set by the stands located in the back part of this space.

空间创意 Space Originality

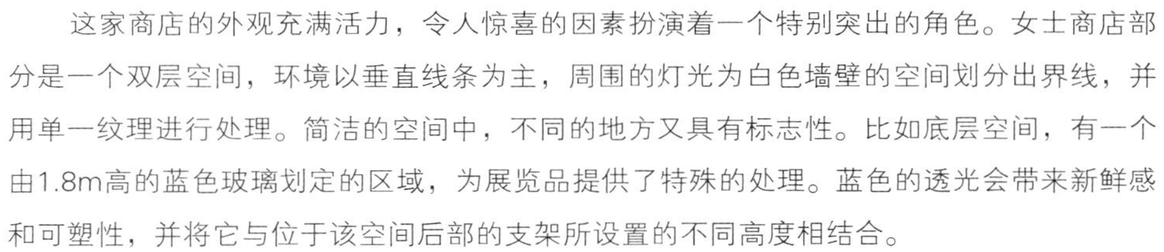

这家商店的外观充满活力，令人惊喜的因素扮演着一个特别突出的角色。女士商店部分是一个双层空间，环境以垂直线条为主，周围的灯光为白色墙壁的空间划分出界线，并用单一纹理进行处理。简洁的空间中，不同的地方又具有标志性。比如底层空间，有一个由1.8m高的蓝色玻璃划定的区域，为展览品提供了特殊的处理。蓝色的透光会带来新鲜感和可塑性，并将它与位于该空间后部的支架所设置的不同高度相结合。

室内陈设
Indoor Furnishings

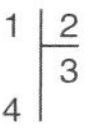

[1] 用家具成就服装的展示

家具由石材、木材和陶瓷等珍贵材料制成，结构简单，却又很引人注目，为服装展览创造了新的路线和安排。通过这种方式，整个空间变得动态化，人在空间中，能保持持续的运动。

[2] 以橱窗为主角之一，给空间创造记忆要点

店里具有许多标志性部分，尤其是橱窗。这是一个透明的空间，由近8m长的墙组成，底部装有8个模块。每当商店橱窗改变时，8个模块就以不同的方式排列，从而完全改变整个空间。橱窗里，是一排排列整齐的同款大衣，曲线的架构使大衣显得笔挺，也能在消费者心中留下深刻印象。

灯光设计 Lighting Design

[1] 以光线划分空间

楼层后部的开放式天窗将自然的光线引入商店，创造了一个与众不同的区域，使其充当更衣室的前厅。

[2] 有针对性的照明，增加光线层次

排式的射灯照明给整体空间开阔明亮之感，壁灯向特定的商品投射光线，更突出商品，也赠与空间多元的深度、浅度，丰富空间次序。

THE CHILDREN'S CLOTHING STORE ENCOUNTERING A SHARK

撞见鲨鱼的童装店

| COCCOLE BIMBI |

Coccole Bimbi was born in Puglia in 1989 as a small children's clothing store. Today, it represents an important reality in the territory and an international representative organization for the "kids fashion system". The new Coccole Bimbi is a luxury store far from the center of the city, in a bigger place to accommodate business management and sales activities, where is an old dismantled repair shop in the production area of Rutigliano (Bari).

Coccole Bimbi原来是一家小型童装店，1989年诞生于意大利普利亚。如今，它代表了本土一个重要的现实状况，也是"儿童时装系统"的一个国际性代表组织。新开的Coccole Bimbi是一家远离市中心的豪华商店，位于Rutigliano（巴里）生产区一个可以容纳商业管理和销售活动的场所中，通过拆除旧的修理店而成。

扫码查看电子书

项目信息
PROJECT INFORMATION

Project Name | Coccole Bimbi

Design Company | Silvio Girolamo Studio

Designer | Silvio Girolamo

Project Location | Bari, Italy

Photographer | Simone Boccuzzi

空间创意
Space Originality

The dark shades of the outside are contrasted with the blue of the sky and the white of the clouds on the “tower”, which are designed to capture the attention of bystanders and to become the background of “social” photos. At the entrance, fifteen concrete cylinders of various heights are placed on the nodes of an imaginary grid to reinterpret the theme of the playground. On the steps, there is a big gray star on the doorstep that invites people to enter.

空间外面的黑色阴影与“塔”上的天空和白云形成鲜明对比，吸引行人的注意，并成为“社交”照片的背景。在入口处，15个不同高度的混凝土圆柱体放置在一个想象中的网格上，重新阐释了运动场的主题。台阶上一颗灰色的大星星，仿佛邀请人们进入。

The interior is a big white box divided into three different areas: shop, logistics, offices. The designer identifies and designs seven different areas. According to different criteria and styles, each area is destined to a specific category of goods, so that represents the epitome of a metropolis, where we can discuss fashion and integrate art, design, fashion and other different languages here.

商店、后勤部、办公室是白色大盒子空间的3个不同部分，设计师确定设计了7个不同的领域，根据不同的标准和风格，每一个领域都有一个特定的商品类别，从而变现出一个都市的缩影。在这里，人们可以谈论时尚，融合艺术、设计和时尚等不同的语言。

室内陈设
Indoor Furnishings

1
2 | 3

[1] 用绿植作为空间引入

一个五颜六色的木质和薄纱结构邀请游客去探索第一个展区，木材、蕨类植物和椰子树形成了一个后现代丛林，还有一个秋千代替藤本植物的空间，妙趣横生。

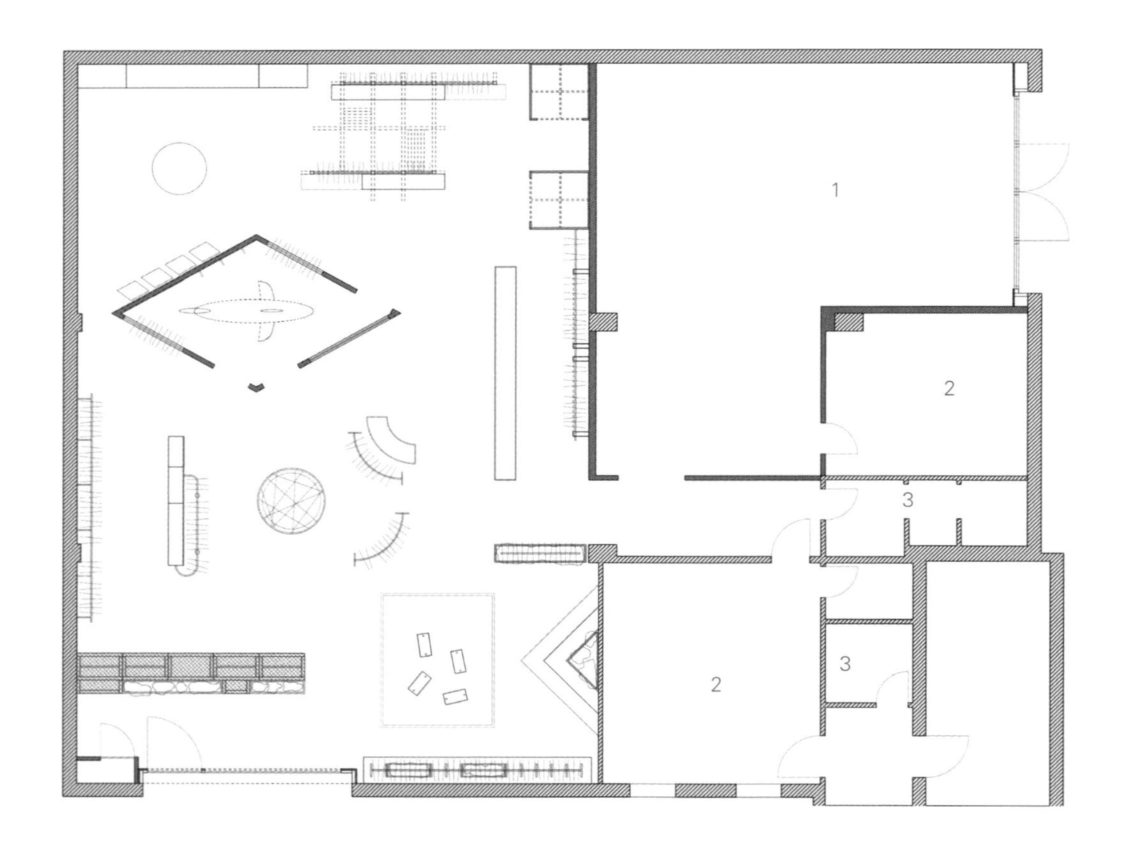

◀ Space Plan

1 Stockroom 储藏室
2 Office 办公室
3 Bathroom 洗手间

[2] 特定空间展示特别的商品

一个菱形的房间，专门用于陈列特别的货品。可活动的货架通过不同形状的三个开口整合到墙壁中，粉红色、蓝色、紫色和黄色强烈的阴影赋予空间激动人心的感觉，使这个区块能立即引起人的关注。

[3] 强化生活空间形象

由木桩构成的“森林”中，小屋的建筑细节和倾斜的屋顶展现出一个工业仓库内典型的山景。试衣间温馨亲切，经过精心设计和特色展示，就像其他空间一样，吸引孩子们的注意力。

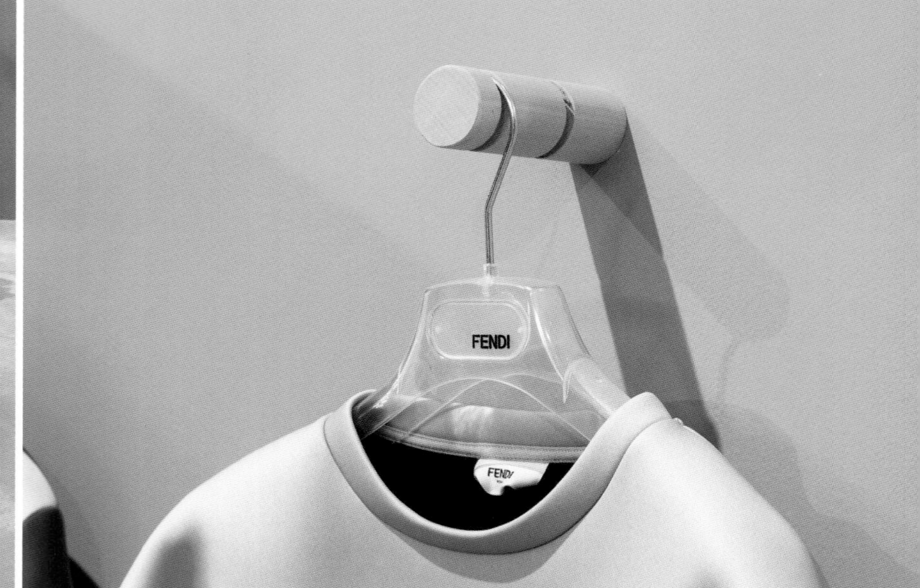

[4] 用玩具抓住孩子的注意力

一条看起来不受干扰地游过所有人头顶的大鲨鱼，非常戏剧化，冲击感十足，能够引发出令人大吃一惊的效应。

灯光设计
Lighting Design

[1] 局部与全局用光的追求效果不同

用生铁材料制成的霓虹灯管、透明玻璃的吊灯、试衣间的线形光，设计师用多样的灯光造型创造局部惊人的灯光效果，或神秘、或温馨、或颇为随性。而空间在整体明亮的光线之中，显得更清新。

[2] 用光“写字”，细心体现呵护

在由旧商店的旧式家具制作而成的新生儿区域，设计师以别样的心思绘出一幅墙上画，即用霓虹灯拼出“BABY”一词，周围摆放上许多短竹，表达出对新生命的关爱。

BRAVISSIMO

COCCOLE BIMBI

THE DREAMY ARCHITECTURE AT THE CORNER, THE CHARMING ROCK STYLE

街角的梦幻建筑，迷人的摇滚风

品牌文化 Brand Culture

| RtA |

RtA was created by Eli Azran and David Rimkoh for promoting sports. As an international designer brand in Los Angeles, two designers inject the meaning of "glamorous rock" into it, considering that the clothing should be a form of expression that is relaxed and chic. RtA also represents a state of mind that makes the wearer more confident in posture and ideas that will never be intimidated by other external substances.

RtA由Eli Azran和David Rimkoh共同打造，以促进运动为目的。由于是洛杉矶的国际设计师品牌，两位设计师为其注入了“迷人摇滚”的含义，并认为服装应该作为一种表达形式，既轻松又别致。RtA还代表一种精神状态，让穿着的人更增强自信的姿态与想法，而这些想法永远不会被其他外在的物质吓倒。

扫码查看电子书

项目信息
PROJECT INFORMATION

Project Name | Road to Awe (RtA)

Design Company | Dan Brunn Architecture

Designer | Monica Heiman

Lighting Design | Dan Brunn Architecture

Project Location | CA, US

Area | 112 ㎡

RtA

R+A

空间创意
Space Originality

The designer creates a complex, dream-like space. There are the geometric precision and the positive and negative dualities, where is a place of meditation.

The store is located at the corner, and the designer completely reshapes its geometry to create a more cohesive, sculptural experience. At the same time, the streamlined boutique proclaims its variably angled black facades to the public, and the storefront is exposed from two sides.

Instead of installing a typical storefront that allows the merchandise to be viewed at a glance, the designer reduces the amount of exterior glass to create a more exclusive and unique atmosphere that allows people on the street can see the clothes at a glance. A large window is set at an angle to face the traffic moving east, while the smaller windows provide views of the interior at the pedestrian scale. Reducing the size of the windows lends an air of exclusivity, offering a particular experience of the brand and encouraging the curious to venture inside to discover the goods. The new "floating" canopy achieves an added sense of mystery in front of the building. A floor-to-ceiling pivot door seamlessly blends with the black exterior when closed, and generously welcomes shoppers when open. There is a constant interplay between open and closed, light and dark, and exposure and concealment.

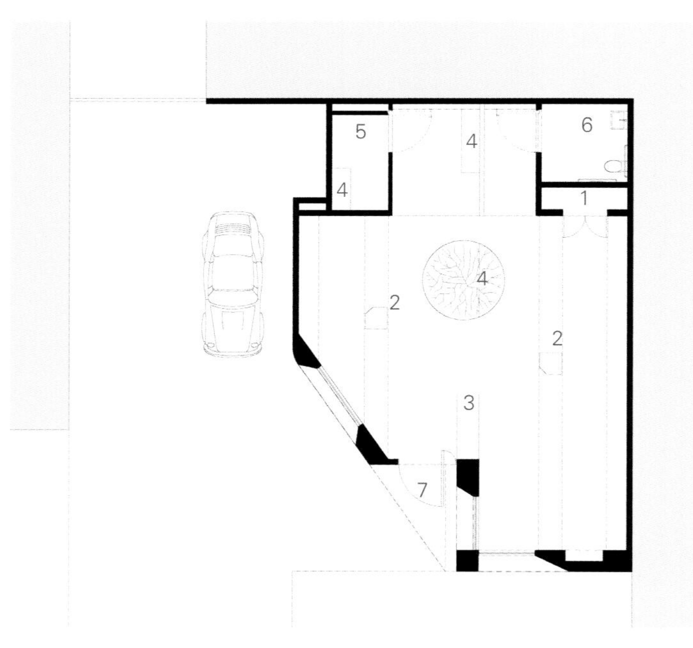

Space Plan

1 Storage 储藏室
2 Spinning Display 旋转展示区
3 POS 收银台
4 Bench 休息凳
5 Fitting Room 试衣间
6 Bathroom 洗手间
7 Entry 入口

设计师创造了一个复杂的梦幻空间。这里有精致的几何型态、正负关系的两面性，是一个让人产生冥思的地方。

店铺位于街角处，设计师重塑了空间的几何形态，创造出具有凝聚力的雕塑感的用户体验，同时流线型的外观拥有面向市民的黑色立面，且从两面都可以看到商店的门面，增加了曝光度。

设计师没有为店面设计典型的门面，而是减少了外部玻璃的数量，创造出可以让街上的游人一眼就能定位到商品的独特体验感。大的橱窗面向东侧的交通空间，与街道之间构成一定的角度；小型的窗户面向街道，为行人展示店铺内的景观。小型的窗户还会给人带来私人化的感受，更好地展示品牌所具有的特性，并激发顾客的好奇心，鼓励他们探索店铺中的商品。建筑物门前“浮动”的遮阳板为空间增添了神秘感。从地板到天花板通高的平开门在关闭时，可以和建筑外立面的黑色外观融合在一起，打开时又会给顾客带来友好亲切的体验。在门的开合之间，光和暗交替作用，展示或隐藏店铺中的商品。

1 | 2 3
4 | 5

[1] 室内造景，代入花园

在10英尺（3.048m）高的空间内，一棵树、草坪和围绕他们的弧形木质长凳构成了一个室内中央花园。其他内部则围绕花园组织，混凝土板、黑色镜子、木质表面、黑色钢梁，创造了前卫时尚的极简主义背景，而成衣展示于两侧，呈现出一个圆形的冥想空间，平静又自然，犹如画龙点睛。

[2] 定制化设计，滑动式陈列

设计师为空间定制了灰色家具、悬挂滚轮轨道等，用纯净的元素打造动态、活泼的销售环境。商品沿着空间延伸，围绕顾客的行走线路布置。服装展示于四个定制的悬挂滚轮轨道上。衣架在空间中滑动，将商品和展示功能整合在设计中，保持空间的丰富性和深度。定制的木箱上安装着试衣镜，旋转的结构展示出用于销售的书籍和配件。

灯光设计

Lighting Design

[1] 用反射光表现品牌本身的气质

树木上方的圆形天窗反射着自然的植被和长凳，并将天光带入整个场所，意在反映亚洲文化在洛杉矶地区的影响力，并展示出品牌本身给人的“敬畏感”。

[2] 在细节处反映品牌的个性

在商店后部的试衣间、等候区和洗手间外部区域，深色饰面下较为昏暗的光线与商店前面部分充足的光线对比鲜明。设计师用一个约3m高的镜子点亮了整个灰色调的空间，给人意料之外的惊喜。木材表面凹槽与灯光一起打造出品牌LOGO的第二个字母即“T”字形的背光标志，仅有的一束光富有强烈的视觉冲撞力，进一步反映了RtA品牌的不羁个性。

EXTREME HIGH-END BLACK, THE UNUSUAL ARTISTIC STYLE OF A SMALL STORE

极致高级黑，小店不凡的艺术格调

Brand Culture

| PODOLYAN |

Collaboration with the fashion designer PODOLYAN started from LOGO design and total rebranding of the company in 2014. Next step was the monopoly store opening in the heart of Kiev.

Its commercial area limited the space provided by the client.Therefore, to emphasize the spaciousness of the place, we decided to produce lightly sophisticated and exquisite furniture and to keep the show window as transparent as possible for a good view from the street. The color pallet of the whole project corresponds to the Podolyan brand: black with light and dark shades of gray. A big show window and specific lighting arrangement which are specially customized gave us an opportunity to design the interior in dark tones, combining a variety of molding textures and corresponding to the color of the brand.

扫码查看电子书

项目信息
PROJECT INFORMATION

Project Name | PODOLYAN Store Project

Design Company | FILD design thinking company

Designer | Dan Vakhrameyev

Project Location | Kiev, Ukraine

Area | 80 m²

Photographer | Roman Pashkovskiy

PODOLYAN

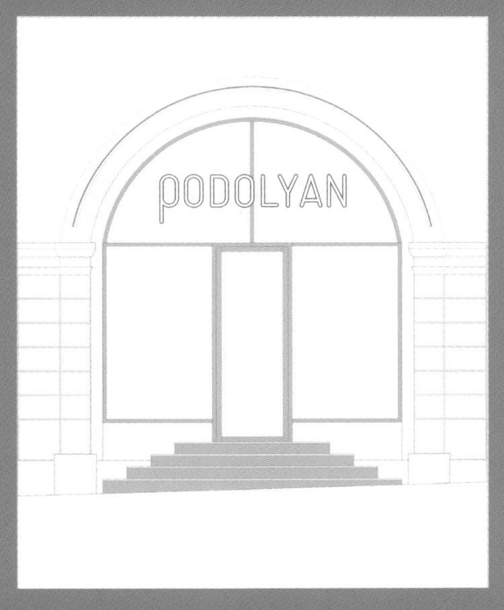

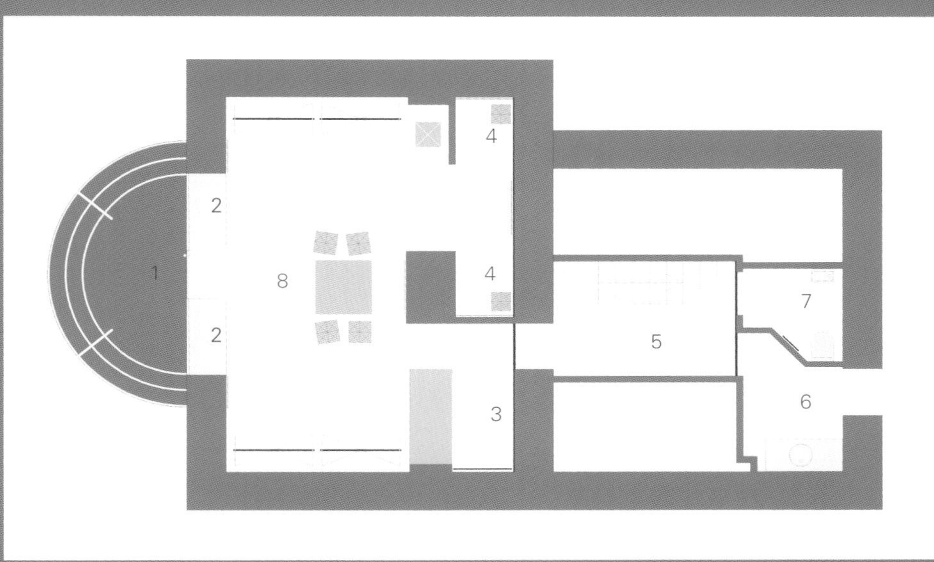

与时装设计师PODOLYAN的合作始于2014年公司的标志设计和品牌重组，然后便是在基辅市中心开设的一家专卖店。

由于客户所提供的空间受到其商业面积的限制，为了强调这个空间的宽敞性，设计师决定生产稍微复杂且精致的家具，并从街上的角度看尽可能保持橱窗的透明度。整体项目的调色板与Podolyan品牌相对应：带有深浅不一灰色调的黑色。特别设置的一个大的展示窗口和详细的灯光照明安排，提供了一个可以结合各种成型的纹理并在深色调下设计的机会，与品牌色调相呼应。

Space Plan

1 Entrance 入口
2 Display Window 橱窗
3 Reception 接待处
4 Fitting Room 试衣间
5 Corridor 过道
6 Kitchen 厨房
7 WC 洗手间
8 Commercial Space 收银台

The concrete tile was customized by SolidHeads specifically for this project, due to a specific intent to escape from standard forms and to create geometrical patterns. Some tile pieces were branded with PODOLYAN LOGO and were spread around the floor in chaotic order to set the brand effect. The black ply-wood decorative panels on the wall emulate the geometric pattern of the floor. Some part of the panels asymmetrically reaches the ceiling and makes a bold stylistic statement. Besides, the decorative panels accomplish a functional purpose hiding the back side door to other facilities. The designer highlighted a stylish atmosphere with a texturized plaster wall treatment also performed in a dark color.

地面混凝土瓷砖是SolidHeads专门为这个项目特制的，表达了其逃离标准形式并创建几何图案这一特定的意图。同时一些瓷砖贴上了PODOLYAN的商标，分散在地板上，设定品牌效应。墙面黑色的装饰胶合面板仿照地板的几何图案，部分面板不对称地直指天花板，显示出其大胆的风格。此外，装饰面板还有一个功能，即将后门隐藏进其他设施里。设计师用深色处理纹理石膏墙的手法来着重强调一种时尚气氛。

室内陈设 Indoor Furnishings

[1] 陈设与布局巧妙融合

室内所有家具、装饰元素和门把手都根据空间的布局定制和设计，与整体格调完美相融。

[2] 用亮点打造空间独有属性

精致家具简约而微妙的线条与深色的室内气氛形成了鲜明的对比，漆成浅灰色家具的薄金属轮廓奇妙地吸引了空间的亮度。架子的结构由连续的线条组成，这些线条在底座或地基上交叉，反映了地砖的几何图案。多功能金属架网格被放置在墙上，用来展示服装和配饰。

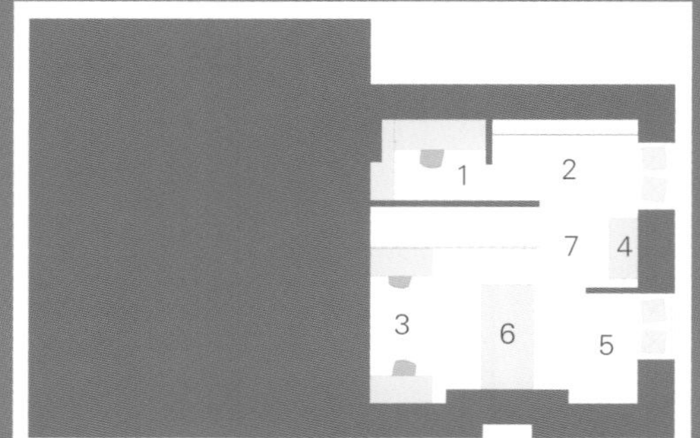

Space Plan

1 Office 办公室
2 Packs 包装区
3 Seamstress 缝合区
4 Ironing 熨烫区
5 Fitting Room 试衣室
6 Tailor 裁剪区
7 Second Floor / Atelier Area 二层 / 工作室

[1] 以灯光营造空间中心点

空间中有一套软垫座椅和一张混凝土表面的桌子，垂直设计的36个爱迪生灯泡漂浮在家具上方，营造出一种舒适的氛围，突出了空间的中心位置。

灯光设计
Lighting Design

[2] 展示品牌，创造吸引力

位于展示窗口的灯，按照其形状，选用柔和的暖光照亮室内。最新的FILD设计系列的12盏灯——代号SO6照亮了衣架上的品牌服装，营造良好的灯光环境，吸引顾客互动、体验和购买。

[3] 多角度层叠设计，形成有效的视觉冲击

灯光的投射会产生不同的明暗效果，多角度层叠的运用可以使服装具有立体感、材质感，并获得展陈所需的照明气氛，突出品牌特质。同时，高亮度突出品牌LOGO，可以提升品牌感。商店的立面和橱窗装饰着一个带有内部照明的品牌标志，入口台阶的线条映衬出橱窗的拱形，效果鲜明。

CARMEN: THE SHOP OF TIME

卡蔓：时光店铺

|CARMEN|

The Hong Kong woman's fashion brand CARMEN, leading the light classic upsurge, was founded in 1997. The founder establishes the core design concept of "showing the fashion with passion, deducing the sensibility with grace", requires the brand not only to pay attention to the appearance of clothing but also to highlight the striking connotation and tolerance. The founder also hopes that every woman in CARMEN is the protagonist of her life to make life bloom brilliantly and will have the courage to live on her way and show their unique charm.

引领经典热潮的香港时尚女装品牌卡蔓（CARMEN）创立于1997年。创始人确立了"以激情展现时尚，以优雅演绎感性"的品牌核心设计理念，要求品牌不仅关注服装的表象，更注重以服装突显消费者耐人寻味的内涵和气度，希望穿着CARMEN的每一位女性都是自己生命的主角，让生命绽放光彩，勇于活出自我，展现自己独具特色的魅力。

项目信息
PROJECT INFORMATION

扫码查看电子书

项目名称｜CARMEN
设计公司｜壹席设计事务所
设 计 师｜胡涛、罗伟伟、罗文良、刘启宏
项目地点｜广东东莞
项目面积｜340㎡
主要材料｜檀香木纹大理石、爵士白（拉丝）大理石、金意陶瓷砖、亚克力、定制型波纹板等
摄 影 师｜刘诚

The

空间创意
Space Originality

The CARMEN flagship store combines with the brand culture and the brand temperament, and space is mainly based on fashion and modernity. The large ceiling installation in the centre of the store enhances the spatial level and also increases the enjoyment of space. In order to create a cultural atmosphere of the commercial space, designers use the abstract painting to balance the sense of distance between business and art.

Designers try to use the cleaner and purer design technique to explore and think the dialogue between people and space and the continuity between people and space, and use a black and white "non-language" scene to create a gestured space stage interpreting the grace, reality and rationality, making space respect the story and continue to write the story.

In the selection of materials, designers focus on the low carbon and economy. The color of materials is mainly in light color, lightening the color of material to create the colorful people and story of the protagonist. A large area of recycled plate enriches the black and white spatial level and details, and the inked color of the local part also shows a degree of relaxation on the rhythm of space.

Space Plan

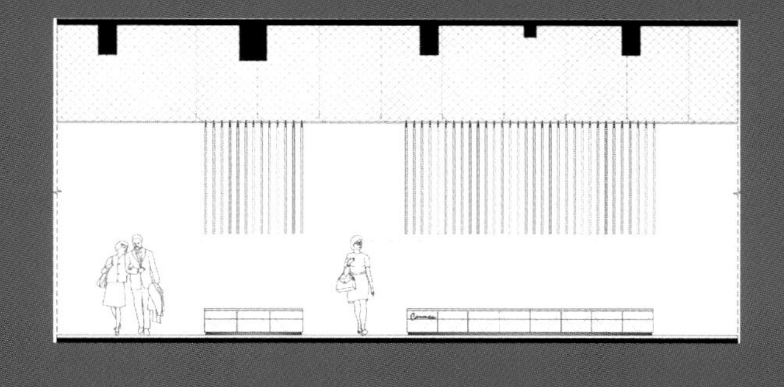

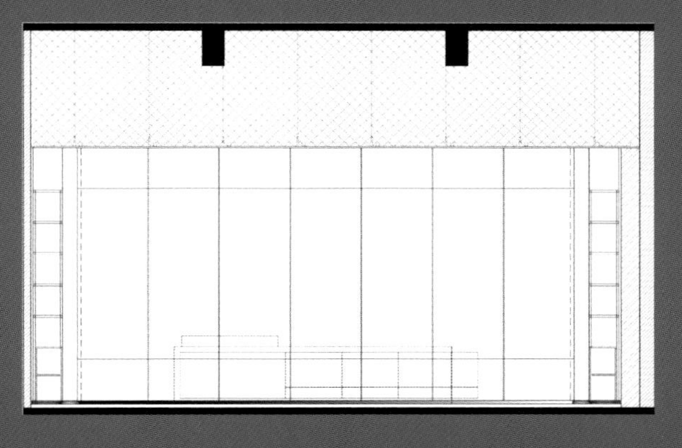

CARMEN旗舰店结合品牌文化与品牌气质，空间主要以时尚现代为主调。店中央的大型天花装置提升了空间层次，也增加了空间的趣味性。为了营造出商业空间的人文气韵，设计师运用抽象画，平衡商业与艺术之间的距离感。

设计师试图以更简洁、更纯粹的手法去探索、思考人与空间的对话、人与空间的共续，并以黑白“无语言”场景为女性空间营造出一个有姿态的空间舞台，演绎从容优雅、真实理性，让空间尊重故事，并续写故事。

在选材上，设计师注重低碳与经济性。材料色调主要以浅色系列为主，淡化材质色调，营造空间主角缤纷的人与故事。大面积的再生板材丰富了黑白的空间层次与细节，局部的浓墨色彩在空间的韵律上也表现得松弛有度。

室内陈设 Indoor Furnishings

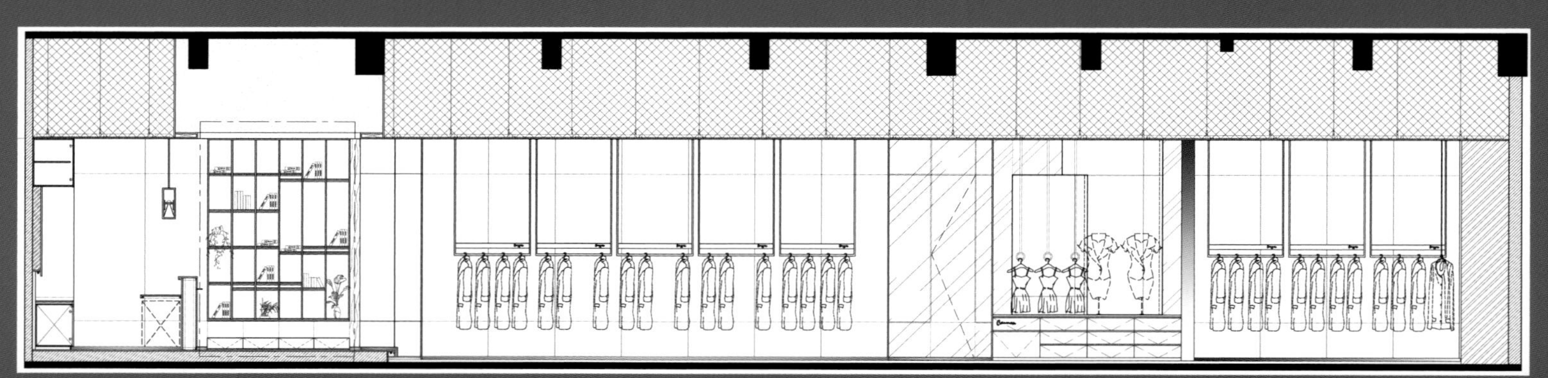

[1] 场景化烘托气场

无论是前台接待区，还是销售区，沙发、单椅、大幅装饰画、灰色渐变的垂直装置以及所展示的商品等构成了一个个小场景，引导购物体验和引发联想。

[2] 连贯布局，增强各功能区互动

表现轻经典的摩登女郎，还有原创甜品品牌THE LUN的休闲区，空间里的一体化陈列带出生活的悠闲姿态，让消费者不仅可以购得满意的服饰，而且也可以享受一个下午的美好时光。

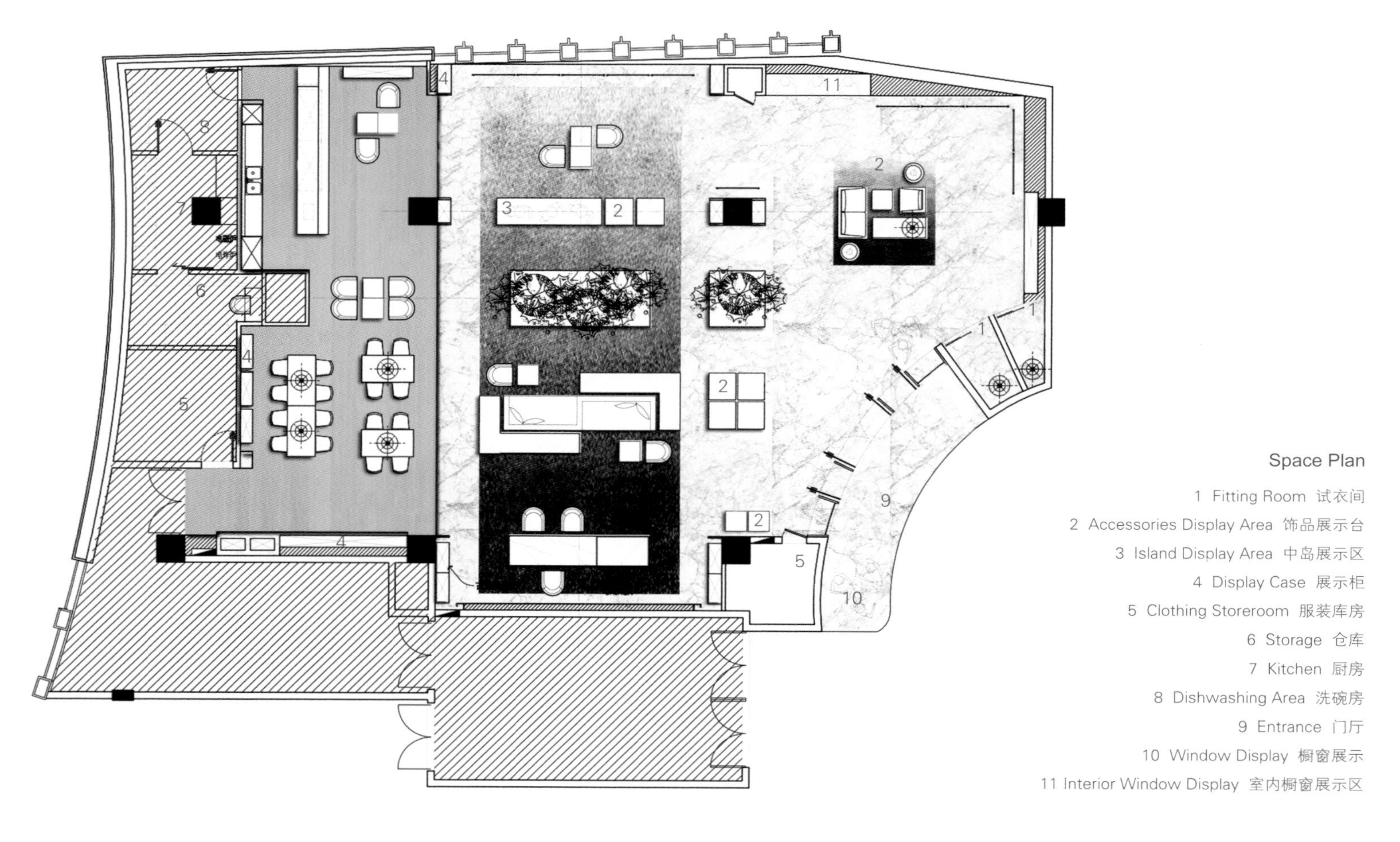

Space Plan

1 Fitting Room 试衣间
2 Accessories Display Area 饰品展示台
3 Island Display Area 中岛展示区
4 Display Case 展示柜
5 Clothing Storeroom 服装库房
6 Storage 仓库
7 Kitchen 厨房
8 Dishwashing Area 洗碗房
9 Entrance 门厅
10 Window Display 橱窗展示
11 Interior Window Display 室内橱窗展示区

The

灯光设计
Lighting Design

不同的灯具配合材质与色彩，彰显商店气质

空间主要选择可调角度的天花灯和分子状枝形吊灯为主要光源，此外还有橱窗区带进来的少量光线，结合大理石和波纹板的质感与色彩，共同打造一个偏冷色调的灰色空间，令空间有一丝高贵的情感。

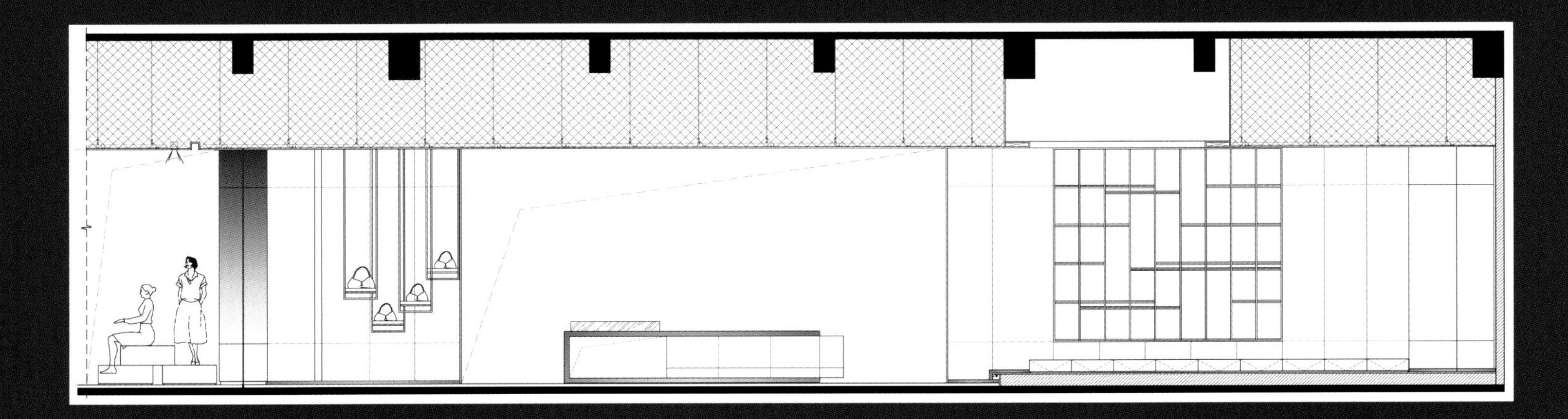

THE MODERN HIGH-END GRAY, THE NEW INTERPRETATION OF FREEDOM AND HUMANITY

现代摩登的高级灰，自在与人文相融的新演绎

Brand Culture

|ZELSOO|

As a Hong Kong designer brand, Zelsoo focuses on tailoring and design, and its target customers are the gentlemen in pursuit of clothing. As a cultural brand which advocates the inner freedom, it takes "inner freedom" and "humanistic concern" as the focus of brand culture, which is exquisite and humane and combines with the art of life to emphasize the feelings of the public on what they wear. Zelsoo combines the aesthetics of freedom into clothing, explores the philosophy of clothing fit for Oriental people, makes clothing become self-cognition, conveys the artistic expression of aesthetics and attempts to guide more demanders to explore the spiritual demands and interpret the attitude of the wearer.

项目信息
PROJECT INFORMATION

扫码查看电子书

项目名称｜执索Zelsoo男装
设计公司｜观觉商业空间设计
设 计 师｜毛邦宇
项目地点｜福建福州
项目面积｜116㎡
主要材料｜真石漆、马赛克、水泥板、水磨石、青铜、熟铁等
摄 影 师｜施凯

执索Zelsoo

执索

执索
ZELSOO

执索
Zelsoo

ZELSOO

空间创意
Space Originality

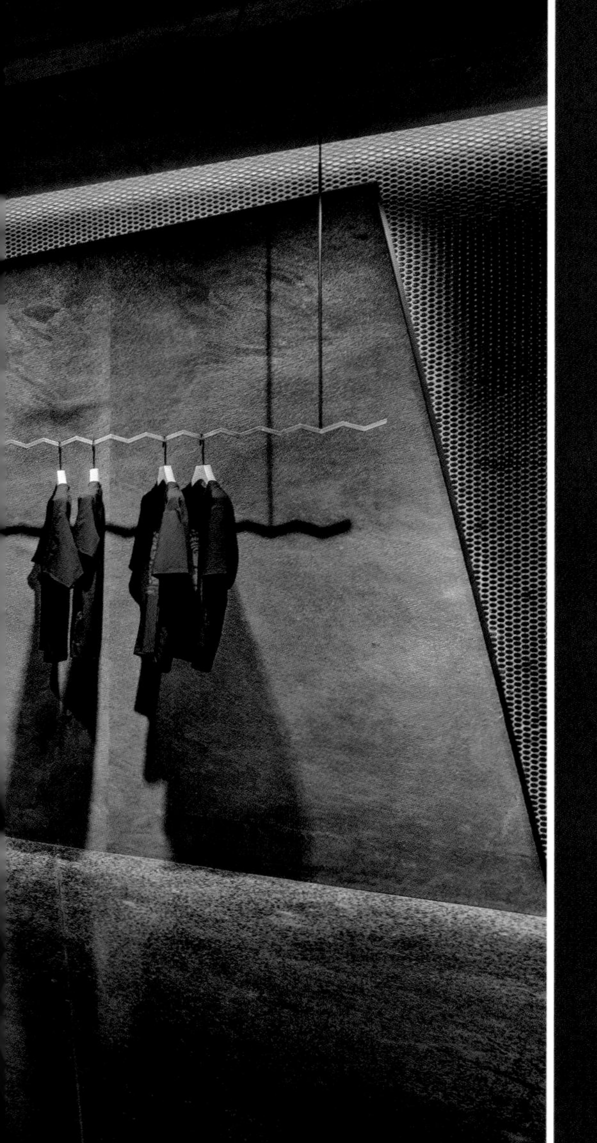

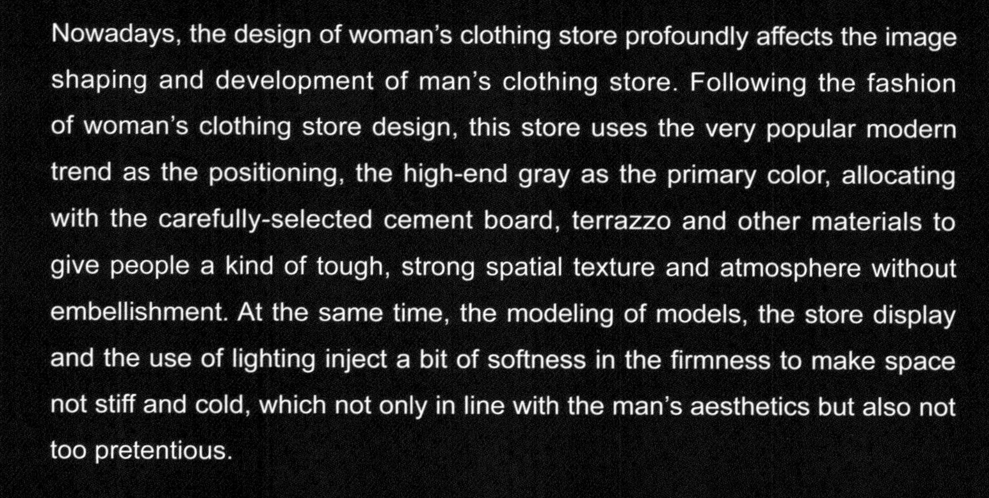

Nowadays, the design of woman's clothing store profoundly affects the image shaping and development of man's clothing store. Following the fashion of woman's clothing store design, this store uses the very popular modern trend as the positioning, the high-end gray as the primary color, allocating with the carefully-selected cement board, terrazzo and other materials to give people a kind of tough, strong spatial texture and atmosphere without embellishment. At the same time, the modeling of models, the store display and the use of lighting inject a bit of softness in the firmness to make space not stiff and cold, which not only in line with the man's aesthetics but also not too pretentious.

在当今时代下，女装店铺的设计深刻影响着男装店铺的形象塑造与发展。这家店铺紧随女装店设计风尚，以时下十分流行的现代摩登风潮为定位，以高级灰为主色调，配合水泥板、水磨石等材料的精心选用，给人一种不事雕琢的感觉，让空间具有坚毅、硬朗的质感和氛围。同时模特的造型、店内的陈列以及灯具的运用，又在刚强中注入一丝柔性，使得空间不显僵硬和冰冷，既符合男士的审美，又不过于矫饰。

室内陈设

Indoor Furnishings

[1] 融入多种材料，以材质烘托色彩

[2] 巧设休息区，吸引顾客逗留

在收银台和试衣间的动线中间，一个黑皮沙发和小圆桌给顾客一个短暂的落脚点。若是逛街试衣累了，小憩一会，喝一杯凉白开，也都体现出店家的贴心。

[3] 一个货架的不同诠释

在一个货架上，不是盲目地用衣物把衣杆填充至过分饱和，而是以颜色为分区，有选择、有节奏地展示衣物，赋予衣物更多的韵律感，方便顾客的欣赏或挑选。

灯光设计
Lighting Design

[1] 隔离亮光，显出空间魅力

在门头顶面，设计师用不规则的木板规划出门口的几何美；店内同样用木质板隔开大量的透亮光源，仅在板与板之间，留出直角的长条形平行光线，给空间适度的光线，用暗调显出空间的高级感，也让试衣间更具诱惑力。

[2] 镂空装扮，加入个性化

背景面板与收银台的镂空给光线的出入留有余地，也在细节之处点亮空间。蒲公英球形灯与镂空有异曲同工之妙，起辅助照明和创意装饰作用。

A LUXURIOUS MONTAGE

一场奢华的蒙太奇

品牌文化 Brand Culture

TOP ZIO

ZIO was created in 1916 and is hailed as the "South Korea first individual men" by the Korea magazine *MOBLIAN*, pursuing the simplicity, fashion and uniqueness. To create a figured man is the rule that ZIO has followed, and ZIO persists in their advocacy of route without any affectation and the complexity.

ZIO创世于1916年，被韩国《MOBLIAN》杂志誉为“韩国第一个性化男装”，追求简单、时髦和别具一格。打造花样男人是ZIO一直遵循的不二法则，始终坚持走自己所倡导的路线，没有一丝矫揉造作，抛弃繁琐。

项目信息
PROJECT INFORMATION

扫码查看电子书

项目名称丨TOP ZIO KOREA 高定男装店
设计公司丨NONG STUDIO
设计师丨汪昶行、朱勤跃、Luca Lanotte[意]
项目地点丨上海
项目面积丨300㎡
主要材料丨黄铜、黑白根大理石、玉纹绿大理石、玉沙玻璃、大花白大理石、木饰面等
摄影师丨汪昶行

空间创意
Space Originality

In the era of information explosion, luxury is endowed

definitions of luxury are already diverse and a mixture of time and space. The design inspiration of TOP ZIO sources from such a luxurious "montage". Designers not only present a kind of spatial intention. Moreover, they seek a mixture of different time and spaces, regional mixture, past and present mixture.

在如今这么一个信息爆炸的时代，奢华也被赋予多重含义。设计师所追求的不单是凡尔赛的宏伟瑰丽，而是沉醉在尼斯的蔚蓝中，也迷失在威尼斯的河道里；他们也渴望一览安第斯山脉的气势磅礴，甚至向往俯瞰非洲草原的万马奔腾，而对于奢华的定义早已是多元的，甚至是时空交错的。TOP ZIO的设计灵感就来源于这样一场奢华的“蒙太奇”，设计师不单只是呈现一种空间意向，而是追求不同的时空交错、地域交流、古今相融。

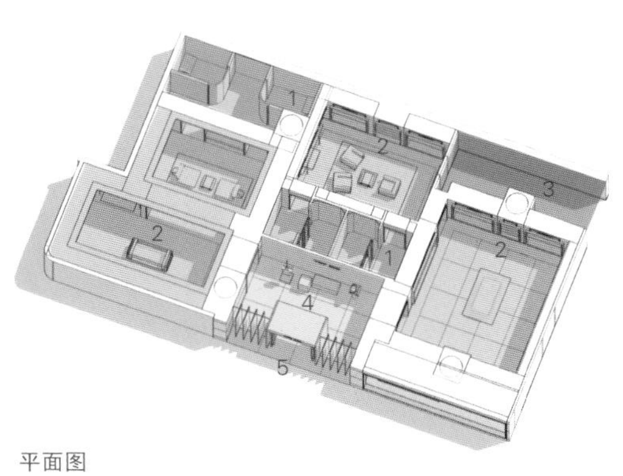

平面图

1 Fitting Room 试衣间
2 Display Area 展示区
3 Storage 储藏室
4 Foyer 玄关
5 Entrance 入口

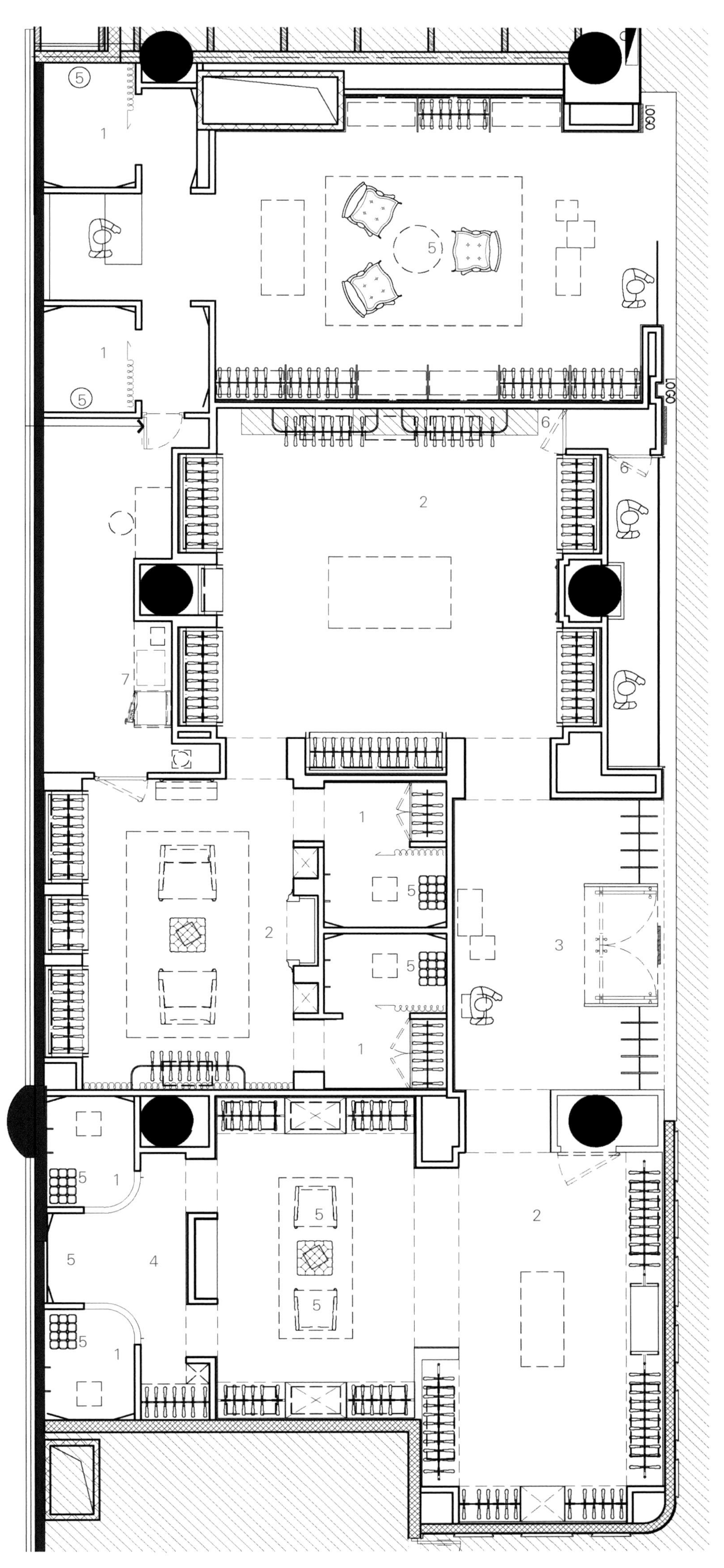

平面图

1 Fitting Room 试衣间
2 Display Area 展厅
3 Main Entrance 主入口
4 Reception 前室
5 Furniture Product 成品家具
6 Blank Door 暗门
7 Utility Room 备用间

室内陈设
Indoor Furnishings

1
2

[1] 串联空间

空间上采用串联的方式连接各个独立的展示空间，从开放的陈列区到半私密的休息配饰区、私密的试衣准备区，最后以开放的辅助陈列收银区收尾，连成一个闭合的购物动线。首先让顾客在心理上感觉仿佛游走于前厅、镜厅、会客室、后庭的交通动线里，其次，不同的空间佐以细微差异从而形成了空间交错感。

[2] 男子气概里的时尚

空间采用格子式布局，构成曲径式通道，给人井然有序的印象。法式烟熏橡木雕花带着复古old-fashion的矜持；外墙与地面的高质感搭配给人时尚的空间体验，配合铮铮然的古铜色金属，展现出男子的阳刚气概。

灯光设计
Lighting Design

[1] 稳重的深邃

入门处设计出神秘的黑色空间，大理石上曲曲折折的线条让空间看起来就像一部旧电影，筒灯照射在西服与雕塑上，指明了故事主角——男士礼服。这里的灯光设计兼具了实用性与艺术性原则，照明的同时增强了展品与装饰物的立体感。

[2] 明暗有致

每一套西服展现在聚光灯下，展现它无声的魅力。以灯带为主营造空间冷静明亮的光线氛围，个别展台的聚光灯给空间灵活的跳跃性，使得整体光效明暗有致，层次丰富，并且主次分明。

[3] 增加展示柜立体感

空间大部分区域做了嵌入式展柜，设计师在展柜四周放置了隐藏光源，增强边框光效使展柜更加立体粗犷。

ATTRIBUTION AND STIMULATION UNDER THE GUIDANCE OF ARTISTIC MAGIC

艺术魔力指引之下的归属与刺激

PLAY LOUNGE

PLAY LOUNGE is a young and dynamic designers shop collection brand. After upgrading the image of the first terminal store, designers establish a friendly and continuable terminal concept for the brand—the "home" of PLAY, and implant a specific temperament which is suitable for the city and geography in every store. "Play" means "game, acting" and "Lounge" means "loitering, relaxing". By combining these two elements, it can express the fashion ideas of a brand that is the relaxation degree and the dynamic and static feeling.

PLAY LOUNGE是一个年轻有活力的设计师集合店品牌，自形象升级后的第一家终端店铺起，设计师为品牌确立下了一个亲和可延续的终端概念——PLAY之“家”，并且在每一家店植入合乎当下城市及地理关系的特定气质。“Play”取“游戏、扮演”之意，“Lounge”则是“闲逛、放松”的意思，二者结合在一起，表现品牌一张一弛、亦动亦静的时尚主张。

项目信息
PROJECT INFORMATION

摄影师 | 吕博
项目面积 | 764 ㎡
项目地点 | 北京
陈列主创 | YOYO高健
参与设计 | 石开云、祝丹阳
主案设计 | 卜天静
设计公司 | 北京吾觉空间装饰设计有限责任公司
项目名称 | PLAY LOUNGE合生汇店

扫码查看电子书

OUNGE

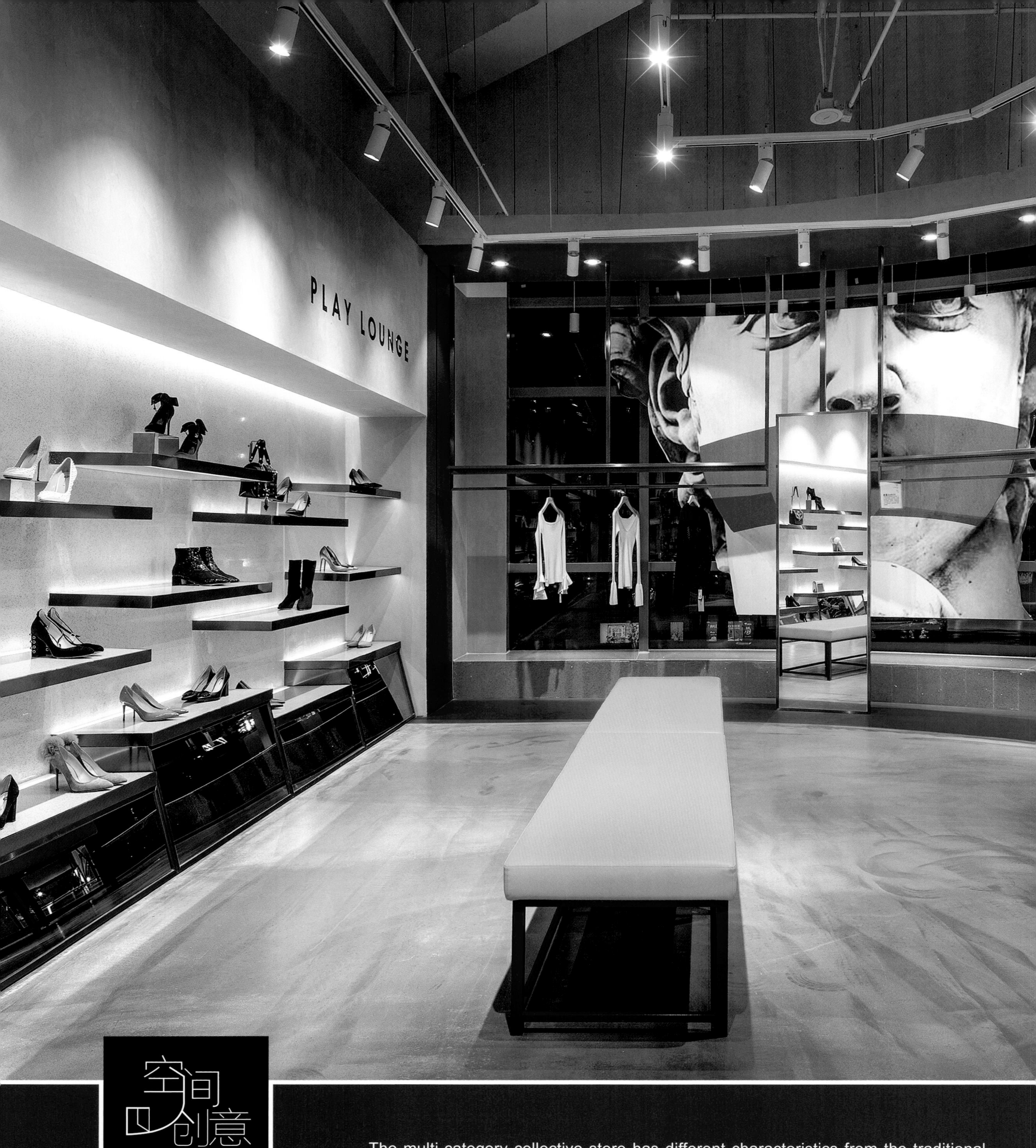

空间创意
Space Originality

The multi-category collective store has different characteristics from the traditional brand stores. According to the various locations, the subtle changes in consumer groups will produce different chemical reactions. This store is near the Today Art Museum and the 22nd Art Plaza, so this particular location has contributed to the birth of the concept of ARTIST HOUSE. Designers hope to add a glimmer of consumer atmosphere with the artistic temperament to this terminal store through the tremendous decorative “house”.

This store includes five designer fashion areas, a coffee bar and a rest area, an accessories area and a beauty salon. There are three blue-purple “ribbons” running

through the store, playing the role of the spatial integration and moving-line guidance. The pillars at the entrance are masqueraded as exhibition information columns of the Art Museum for displaying the information on designers brand or activities. The security gate next to the information column adds a touch of drama to the way of entering the ARTIST HOUSE.

多品类集合店具有不同于传统品牌店的特质，针对不同选址的所在地域，消费群体的微妙变化会产生不同的化学反应。这家店铺所在商区毗邻今日美术馆以及22th艺术街区，特殊的地理位置促成了ARTIST HOUSE概念的诞生，设计师希望通过大艺术“家”给这一家终端店铺增加一丝具有艺术气质的消费氛围。

卖场内共包含5个设计师服装区，此外还有一个咖啡吧及休息区、鞋包配饰区以及美容沙龙。三条蓝紫色的“丝带”贯穿全场，起到空间整合以及动线引导的作用。入口处的柱子被伪装成艺术馆的展览信息柱，用于展示设计师品牌信息或者活动信息。信息柱旁边的安检门为ARTIST HOUSE的进入方式增加了一丝戏剧性。

PLAY LOUNGE

[1] **艺术摆设与蒸汽波相辅相成，赋予店铺“艺术馆”的魅力**

作为迷幻系列的第四部曲，这个空间虽然低调了许多，但设计师又加入了时下风靡时尚圈的蒸汽波（VAPORWAVE），使经典石膏更趣味化。而大卫、维纳斯等经典雕塑形象以当代的手法融入到整个空间中，Casa Gaia创新、艳丽闪耀的欧式拼布沙发等，也增强消费者与主题空间的互动感。

[2] **化直为圆，打造一站式服务**

在有边界的固定空间里，设计师融入了美容沙龙和休息区等，化解有限空间带给访客的局限感。蓝色的维纳斯以及伪装成检票口的咖啡吧台，为消费者在店内的购物体验拉开序幕。消费者进入空间，除了能找到美丽的新衣和有趣的潮品，也能找到享受一杯下午茶的落脚之处。利落的货架上，放上精心挑选过的服饰，让商品在不经意间便成为访客的凝视之物，也让空间生出归属感，化身为每一位消费者放松的场所。

PAY
OUNGE

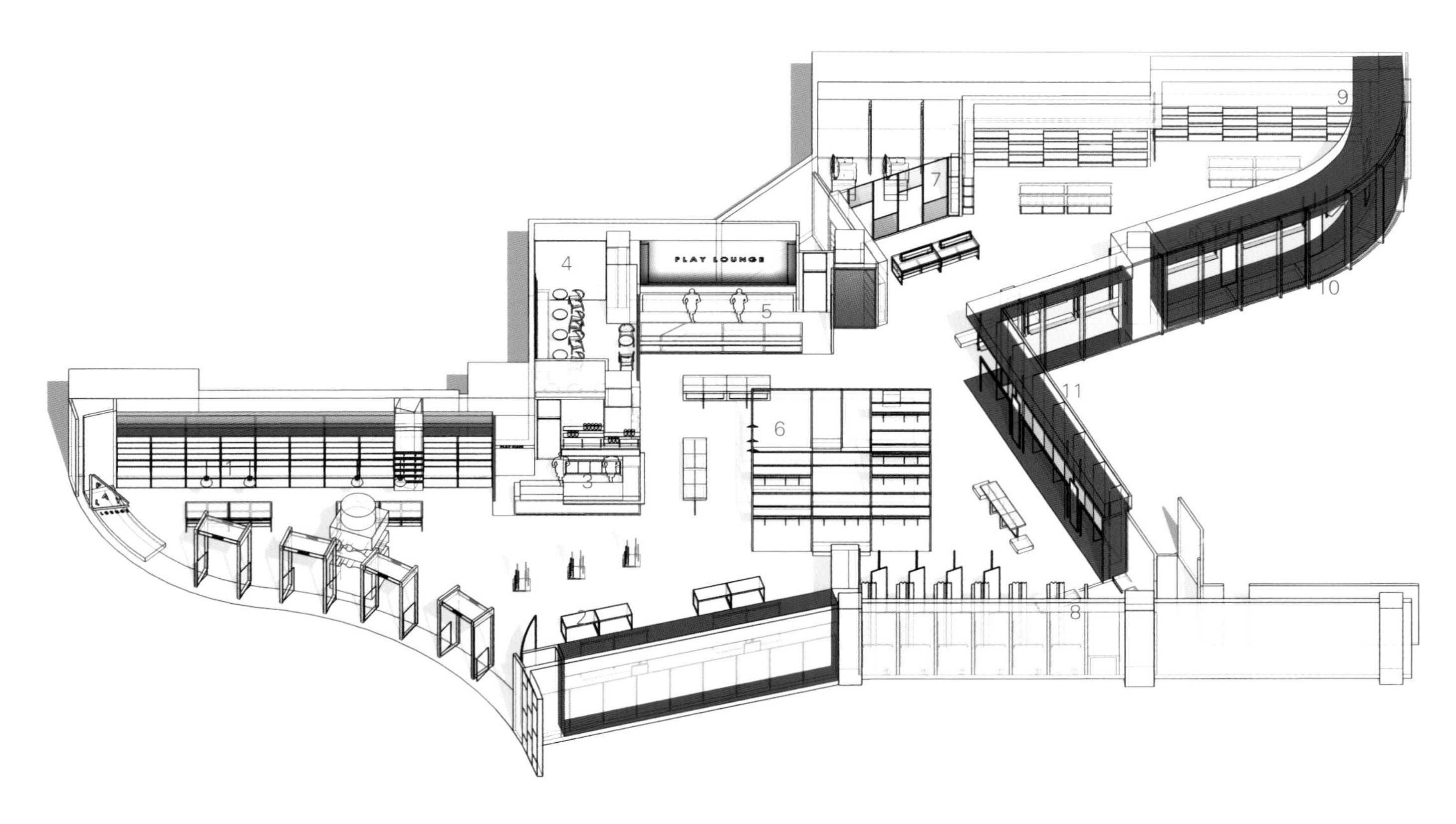

Space Plan

1 Zone B B区域
2 Zone A A区域
3 Cafe Bar 咖啡吧台区
4 Cafe 咖啡区
5 Reception 接待处
6 Zone C C区域
7 Salon 沙龙
8 Fitting 试衣间
9 Accessories 饰品间
10 Zone E E区域
11 Zone D D区域

PLAY

PLAY LOUNGE
LOUNGE
PLAY SALON

Y LOUNGE
PLAY SALON

灯光设计
Lighting Design

1 | 2
3

[1] 用线性光源引出店内的酷炫气质

幻彩玻璃是自设计师打造第一家店 “THE RHYTEM OF FANTASY（迷幻乐章）” 之后，一直采用的作为品牌终端的标志性元素。这次的设计中，设计师同样沿用了这一元素，并加入了蓝色的线性光源，结合汽车表层工艺开发的幻彩银金属板，降低了甜腻，增加了一抹神秘与未来感。

[2] 用光彩突出品牌标志

橱窗式陈列的LOGO区，用光与材质的结合，营造出渐变反光的色调，表现摩登时尚的品牌追求。

[3] 因功能而异，塑造不同的灯光效果

店内色彩纷呈，设计师针对不同的区域，选用不同的灯具，或是小台灯，或是嵌入式天花射灯，或是旋转天花射灯，协调而富于休闲气息，循序渐进，引导消费者步步深入，感悟这个店铺的独特。

THE FASHIONABLE "UPSIDE DOWN"

时尚『逆世界』

| SO WHAT |

What is the brand spirit of SO WHAT? The owner of SO WHAT not only wants to express a self-breakthrough, undefined style but also should have the elegant and witty temperament through the retail space.

什么是SO WHAT的品牌精神？SO WHAT的业主希望通过零售空间表达一种突破自我、不被定义的风格，同时又要具备优雅俏皮的气质。

项目信息
PROJECT INFORMATION

扫码查看电子书

项目名称 | SO WHAT成都时尚店
设计公司 | Circle Studio Shanghai / 上海有寻建筑设计事务所
设 计 师 | 汪琳、尹军、余锦星、王丹
项目地点 | 四川成都
项目面积 | 300 ㎡
主要材料 | 艺术水泥、电镀不锈钢杆件、磨砂白安全玻璃、超白玻璃等
摄 影 师 | 鲁鲁西

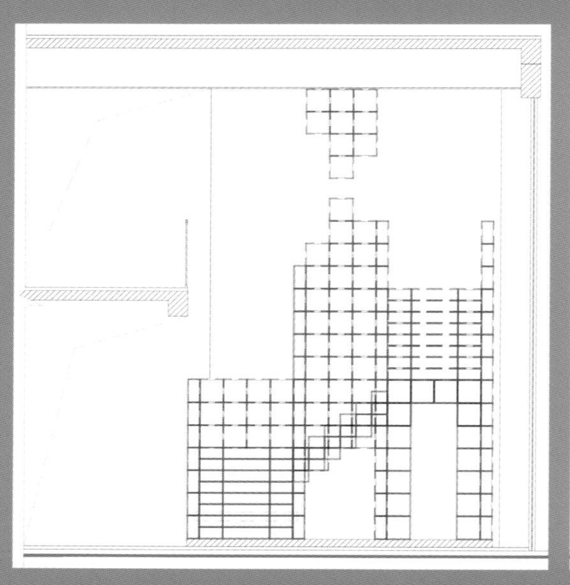

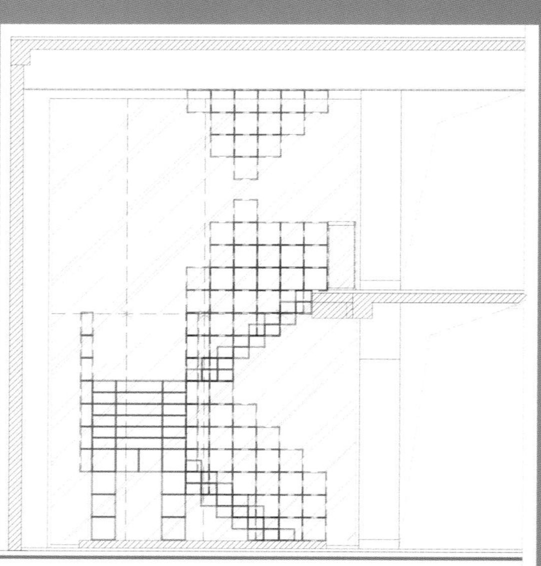

Elevation Stair 1

Elevation Stair 2

1 | 2 | 3

How to express this kind of spiritual character in space design? With this question mark, we begin to explore the concept of design. After analyzing the basic data, a clear intention emerged: *Upside Down*. This is a film about two worlds that are mutually reversed, people are living in the twin world and having a parallel life, and they are unable to touch each other. Finally, with the reversal love of the leading actor and actress, two worlds are linked. Inspired by two storefronts of the SO WHAT shop, designers immediately associate with this fascinating subject. The concept of "Upside Down" can perfectly solve the physical state of the respective separation of the two floors, which forms a tightly bonded space and achieves a kind of space tension that can not be achieved by the simple repeated design.

在空间设计上如何表达这种精神特质？带着这个问号，设计师开始了设计概念的探索。在进行基础资料分析之后，一个清晰的意向浮现在设计师的脑海——《逆世界》。这是一部电影，它讲述的是互为颠倒的两个世界，人们在这个双生世界里平行生活，彼此无法触及，但最终男女主角凭着逆转天地的爱恋，将两个世界连接起来。正是受到SO WHAT店铺临街两层楼面的启发，让设计师联想到这个迷人的主题。“逆世界”的概念可以完美地解决两个楼层各自分离的物理状态，从而形成一个紧密黏合的空间，并达到一种楼层简单重复设计所无法企及的空间张力。

[1] “逆世界”的模块

在这样一个镜像的空间里，设计师只希望添加一种元素——一个边长36cm的立方体框架，既可以模拟“逆世界”的城市结构，又可以陈列商品。一楼的道具从地面升起，二楼的道具从吊顶悬挂而下。同时，将一楼地坪和吊顶的色彩与二楼进行镜像设计，两个楼层就自然而然形成逆转的双生世界，二楼再用倒挂的模特来强调这种空间感。

[2] 多功能道具

模块化的立方体框架像乐高积木一样，通过拼接、堆叠形成悬挂衣服、搁置配饰的道具。同时，为框架设计的可动小构件，搭配玻璃板，可将配饰陈列在不同的框架上。店主只需稍花心思，便可每天都玩出新的陈列。

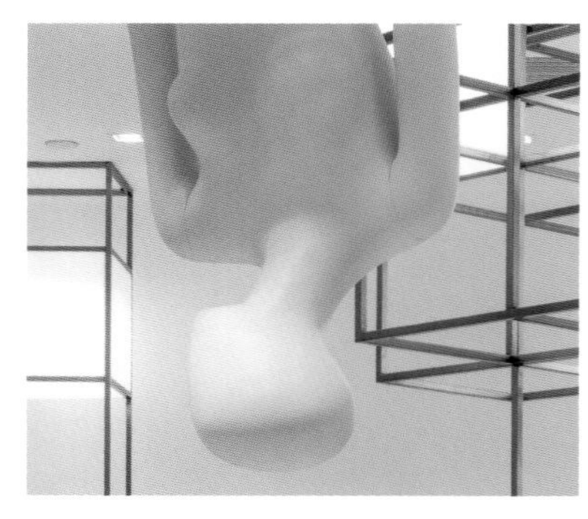

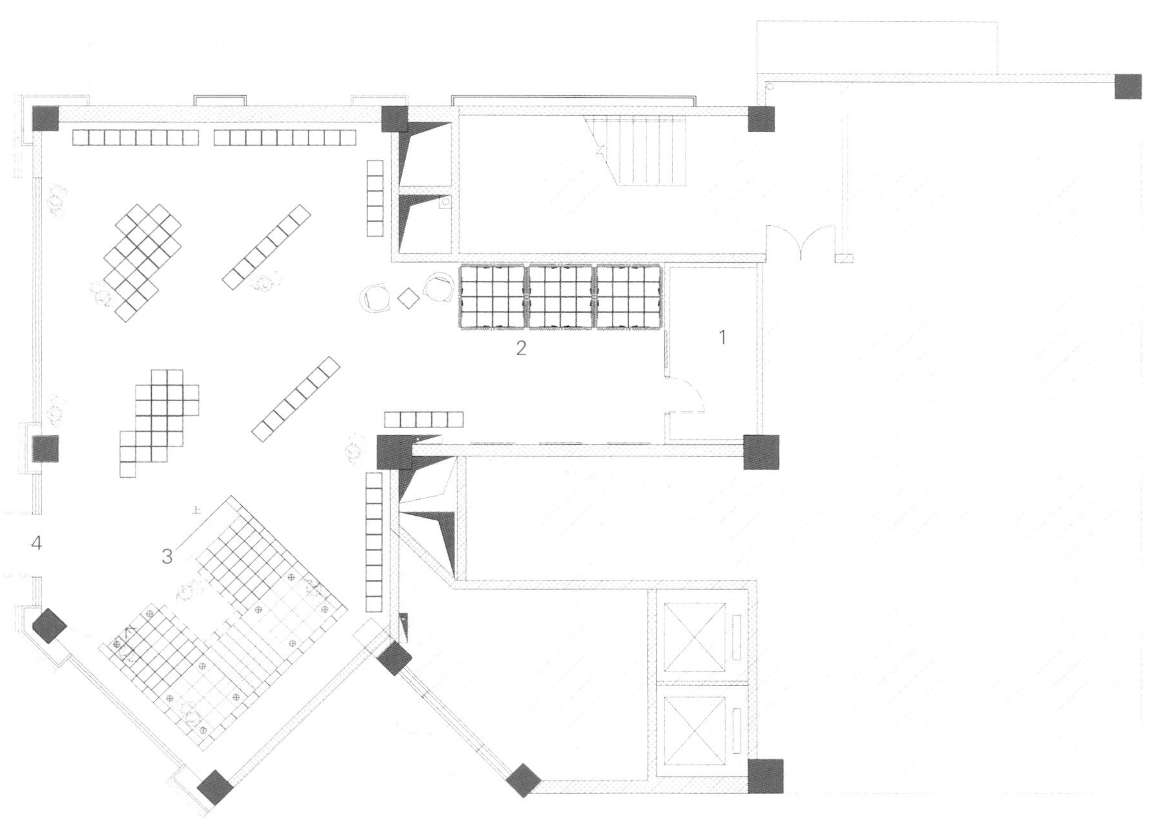

1st floor plan

1 Storage 储藏室

2 Fitting Room 试衣间

3 Staircase 楼梯

4 Entrance 入口

[1] 极简主义空间设计

店铺设计采用极简主义的手法，没有额外的装饰和过度的色彩，统一的陈列元素将商品突显成空间的主角，人们漫步其中感受到的更多是一种艺廊的体验，而非常规的服装店。

[2] 以装饰物来点题

为了突出“逆世界”的空间主题，设计师在天花板上做了许多金属模块装置，以立方体的灯光装置作为修饰，提升了空间的整体格调。

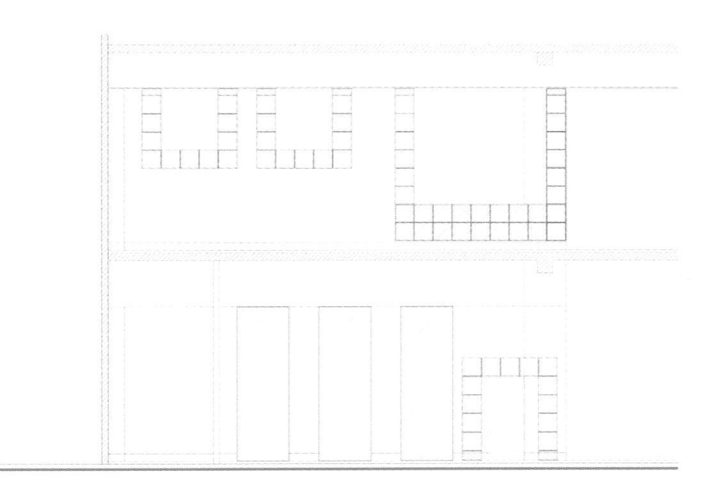

Section A

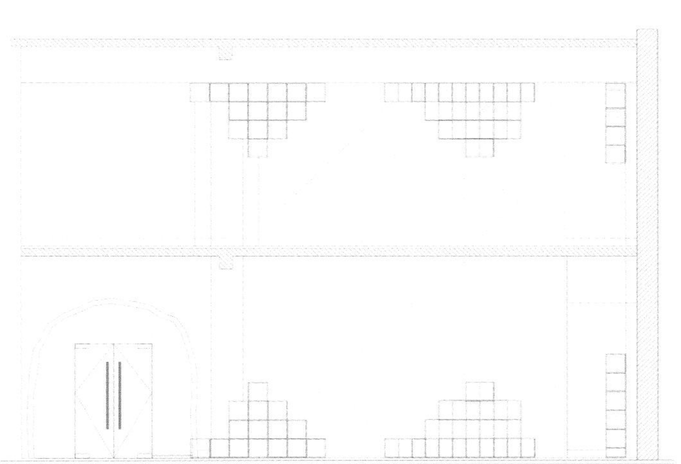

Section B

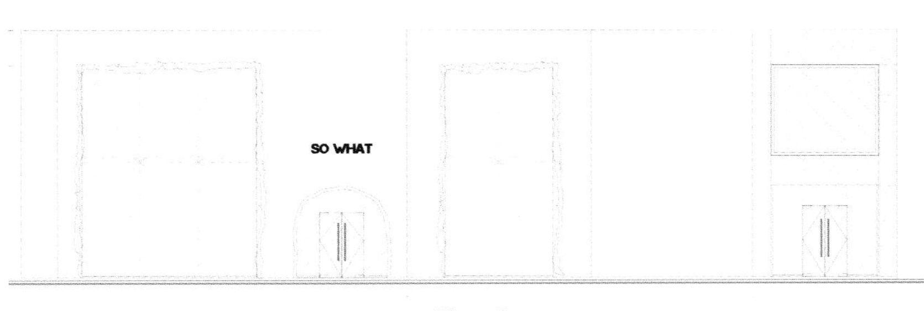

Elevation

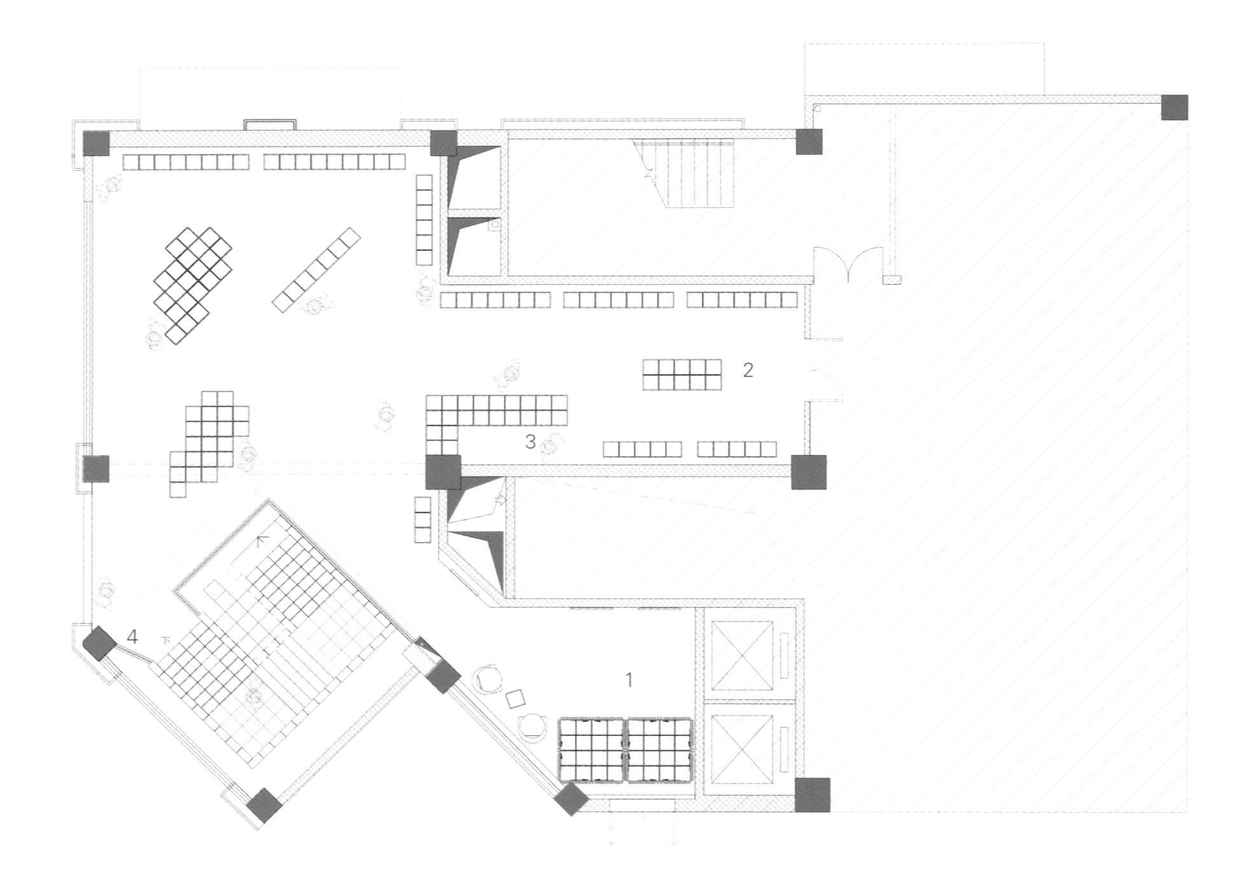

2nd floor plan

1 Fitting Room 试衣间

2 Entrance 入口

3 Checkout Counter 收银台

4 Staircase 楼梯

SO WHAT

THE MODULAR "LILONG"

购物空间的模块『里弄』

Brand culture

| CHUANG × YI |

Chuang x Yi is a fashion concept integration store that has been created especially by the VILLAGE to support the Chinese creative people and the original local forces. Chuang x Yi brings the most influential works of independent Chinese designers to the visionary consumers. Chuang x Yi's co-designers are eclectic, and their design styles are very different. However, they all capture the essence of the contemporary fashion in China from different perspectives and use their unique, pioneering and innovative concepts to express it, bringing a breeze to Chinese customers.

创x奕是奕欧购物村为扶持中国创意人、支持本土原创力量而特别开创的时尚概念集成店。创x奕为眼光独到的消费者们带来目前最具影响力的中国独立设计师作品。创x奕的合作设计师们不拘一格，设计风格大相径庭，但都从不同的角度抓住了中国当代时尚的精髓，并以其独到前卫的创新理念表达，为中国顾客带来一股清风。

项目信息
PROJECT INFORMATION

扫码查看电子书

项目名称 | 创×奕时尚概念集成店
设计公司 | 芝作室
项目团队 | Christina Luk, Marcello Chiado Rana, Alba Beroiz Blazquez
室内灯光设计 | LUKSTUDIO
展示家具与定制灯具设计 | 体物设计
项目地点 | 上海
项目面积 | 150 ㎡
摄影师 | Dirk Weiblen

[创×奕]
MADE BY CHINA

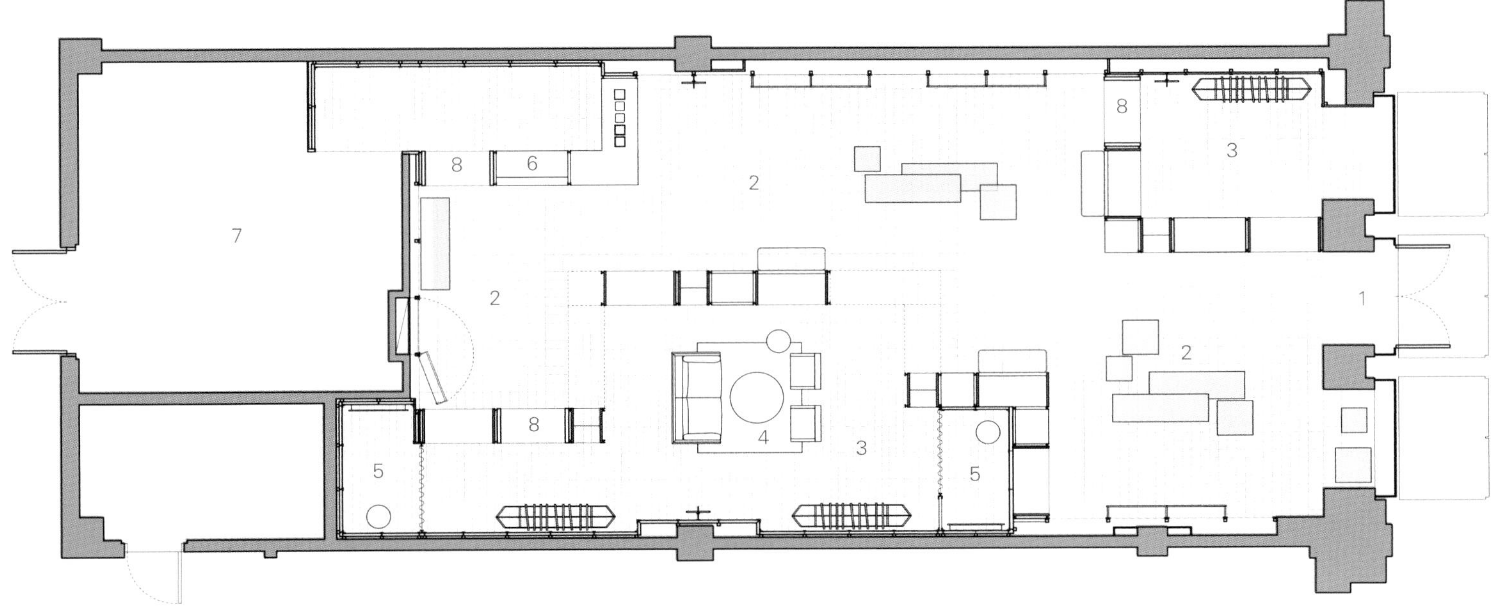

Space Plan

1 Main Entrance 主入口
2 "Lane Way" "弄"
3 Lanehouses "弄屋"
4 Lounge 休息区
5 Fitting Room 试衣间
6 Checkout Counter 收银台
7 Storage 储藏室
8 Shikumen 石库门

[创×奕]
MADE BY CHINA

空间创意
Space Originality

As a kind of urban architectural typology, the spatial limitation of Lilong excites the creativity of people who inhabit here, and you can always find the interesting spatial application and the rich details texture in the meandering lanes. The resulting blur interweaves the private and public space as well as the residential and commercial space, and its evolving organic systems compose the unique urban landscape of Shanghai. In this design, the designer tries to interpret the relationship between a "lane" and three "lanehouses", and divides the different areas of the shop through this concept: the display area, the waiting lounge, the dressing rooms, the main cashier counter and a service area.

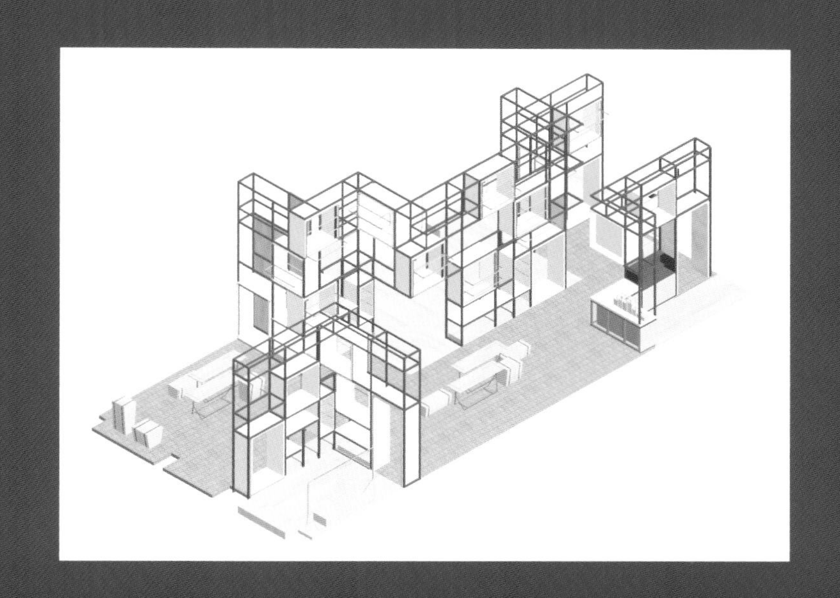

“里弄”作为一种都市建筑型态，空间的局限性激发出栖息者的创造力，延绵的窄巷内总能发现有趣的空间应用和丰富的细节肌理。这种模糊交错着私人与公共、居住与营商，不断自我衍变的有机体系构成了上海独特的城市景观。在这个设计中，设计师试图阐释一条“弄”与三座“弄屋”的关系，并通过这个概念来划分店铺的不同区域：展示空间、等候区、试衣间、收银台以及后勤区。

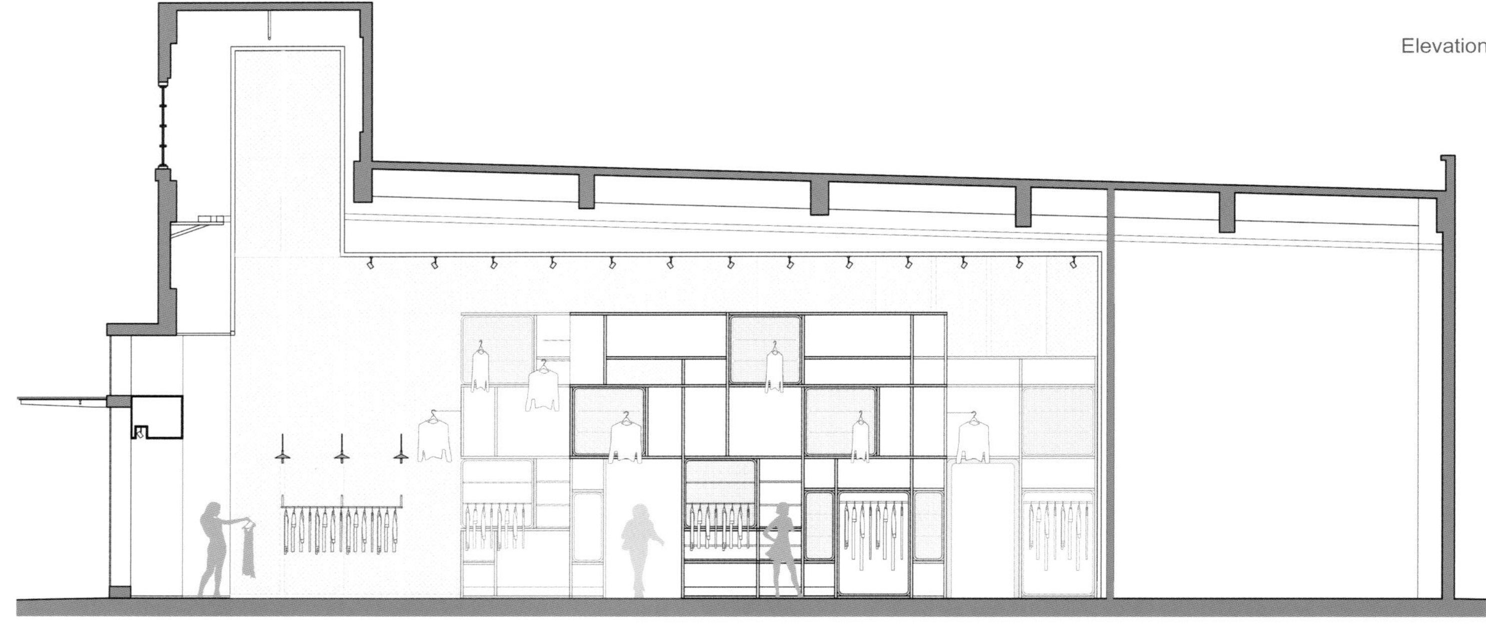

[1] **老建筑的新设计**

铺石与木地板界定空间的同时，开放式的展示形式使得“弄”与“弄屋”之间保有视觉上的延续。多层次的相关元素堆叠成紧密的整体，兼具灵活性与秩序感的体验恰如“里弄”生活的缩影。

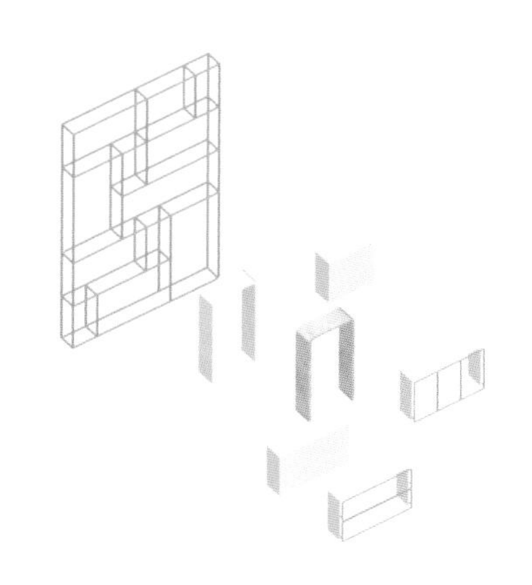

[2] “弄”主题

三座“弄屋”的结构由模块化的金属框及活动单元组合而成，便于店铺换址时拆卸重装。单元的设计取材于“里弄”中丰富的元素：入口单元可看到经典的“石库门”弯角；上海市井标志性的外伸晾衣架化身成服装展示架；旧式凉椅的竹编被用作单元的分隔；可移动的定制家具在“弄”内提供灵活的商品展示，由黄铜和木材做成，让人联想起弄里的长椅桌凳。

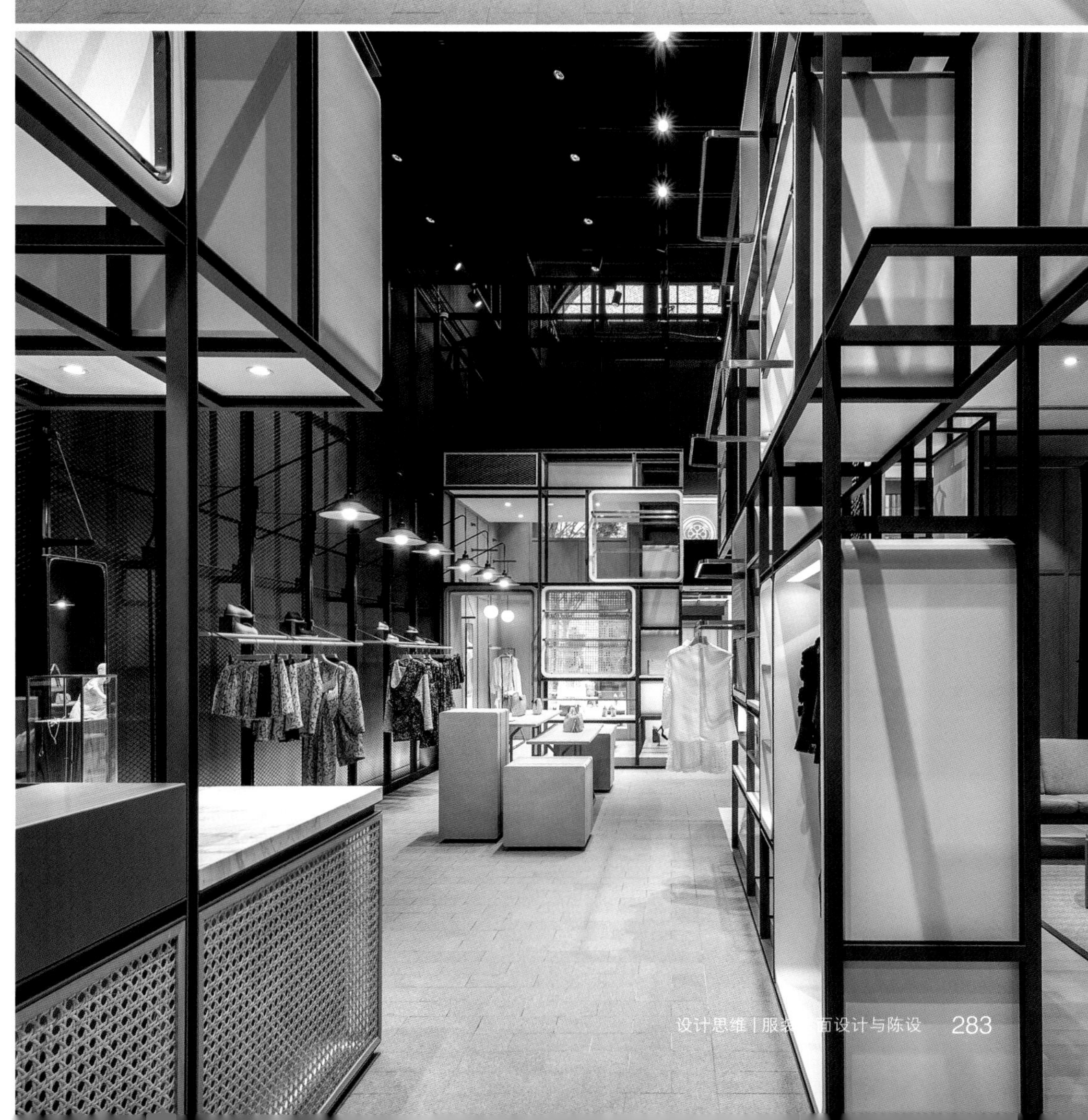

灯光设计

Lighting Design

1 | 2 / 3

[1] 整体灯光

天花板以白光、黄光的射灯为主，给人明亮的空间感受。展示柜增加了光源以突显产品的形象。

[2] 灯具的装饰性作用

空间整体气氛因木色与暖光灯的搭配渲染得十分温馨优雅，灯具也非常简单，一盏盏灵动的球形灯点缀得空间更加讨喜和可爱，而路灯造型的吊灯给空间增加了文艺气息。

THE INTERNATIONAL FASHION ROOTED IN THE CHINESE CULTURE

根植于中国文化的国际时尚

| HOUSE OF GRACE CHEN |

When Grace Chen, an haute couture designer, first approached Kokaistudios and set out her vision of transforming a historic villa into a house of the namesake brand, the Kokaistudios design team was immediately attracted by this idea. Grace Chen's fashion style, the high-level custom-tailored fashion designer, is rooted in Chinese culture but has an international spirit and a personalized fashion atmosphere. The brand is very concerned about the handicrafts and details, with the elegant and meaningful temperament and the exquisite craftsmanship.

当高级定制服装设计师Grace Chen初次接触设计公司Kokaistudios，并阐述她有将一幢历史别墅改造为同名品牌之家的愿景时，Kokaistudios设计团队立即被这一想法吸引了。Grace Chen的服装风格植根于中国文化，但又包容国际化精神，具有个性化的时尚气息。品牌本身十分关注手工艺与细节，具有优雅而隽永的气质与精湛的工艺。

项目信息
PROJECT INFORMATION

扫码查看电子书

摄影师 I Seth Powers
项目面积 I 860 ㎡
项目地点 I 上海
灯光设计 I Kokai Studios
设计团队 I 郑泳、陈颖、陶韦、顾瑞雪、黄冰冰
设计责任人 I Andrea Destefanis、Filippo Gabbiani
设计公司 I Kokai Studios
项目名称 I 瑰丝·陈花园

GRACE CHEN

空间创意
Space Originality

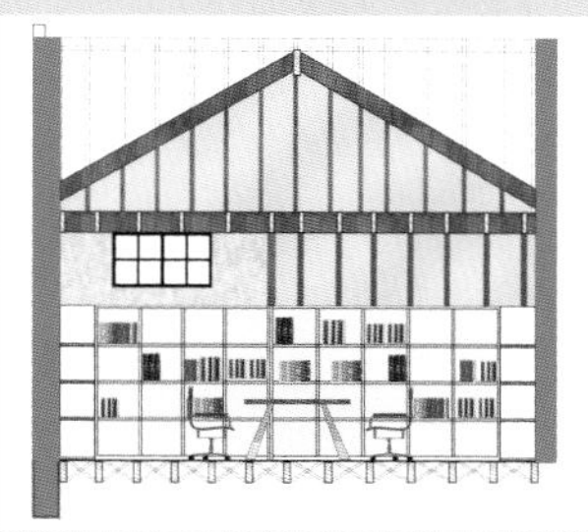

When designing this classical villa, designers inject a sense of modernity so that can make the project more perfect and have a sense of the times. The functional area includes a showroom, gallery, fitting rooms, offices, a library, a show kitchen as well as a VIP suite for a full lifestyle experience. In addition to the architectural renovation and the interior design of the villa, the design team redesigned the garden, including the glass showroom. The garden area is able to host the events or the small fashion shows which are suitable to the needs of customers.

设计师在改造这个气质古典的别墅时，注入了一些现代感，从而使得项目更完善且具有时代感。功能区包括展示厅、艺廊、试衣间、办公室、阅览室、厨房以及提供完整体验的VIP套房。除了别墅的建筑改造和室内设计之外，Kokaistudios设计团队重新设计了包括玻璃展示厅在内的花园，花园庭院的空间既可以举办活动又可用作走秀场，十分贴合客户的需求。

室内陈设
Indoor Furnishings

平面图

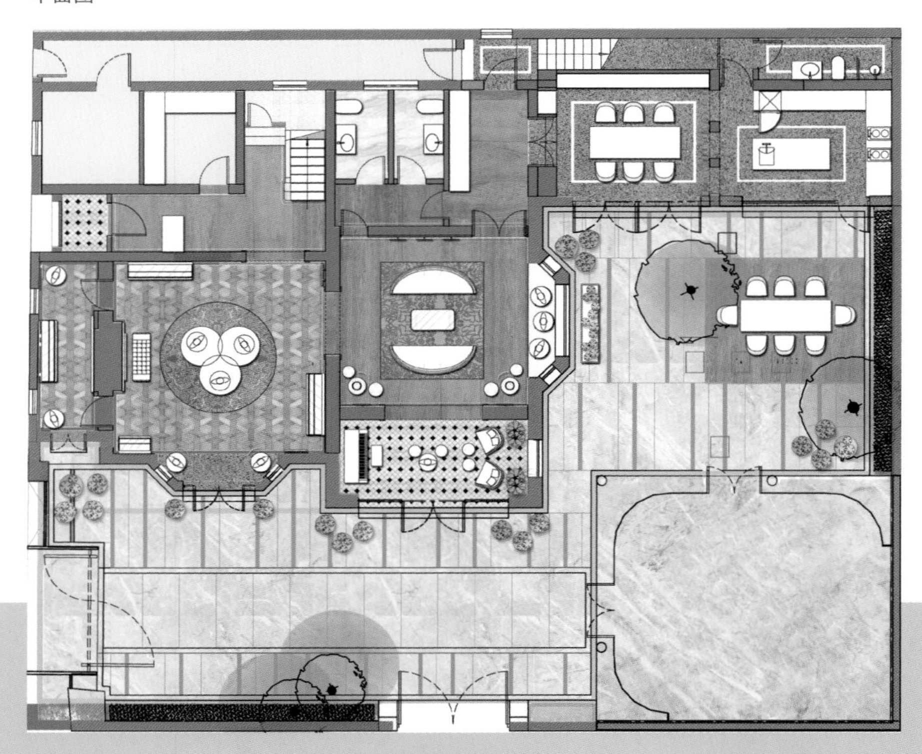

[1] 空间布局

空间主题色彩选择了自然柔和的色系，设计师使用Marmorino涂料为空间营造出温馨感，一方面打造出一个舒适环境，另一方面能很好地衬托华美的礼服产品。一楼展示区与会客区的墙面布满定制古董镜，洋溢着20世纪30年代的装饰主义风格，二楼的主要功能为试衣间，其中两间在风格上延续了女性化的温柔感觉，另一间则比较中性，为日后男装的面世做准备。

[2] 立体展示

空间整体光效以暖光为主，呼应了温馨感。采用直线式陈列，服装主要展示在空间四周。模特的摆放比较灵活，展示区以陈列式布局为主。展示的服装不在多而在于优，步入后环顾一周便能在最短的时间里找到自己心仪的款式。

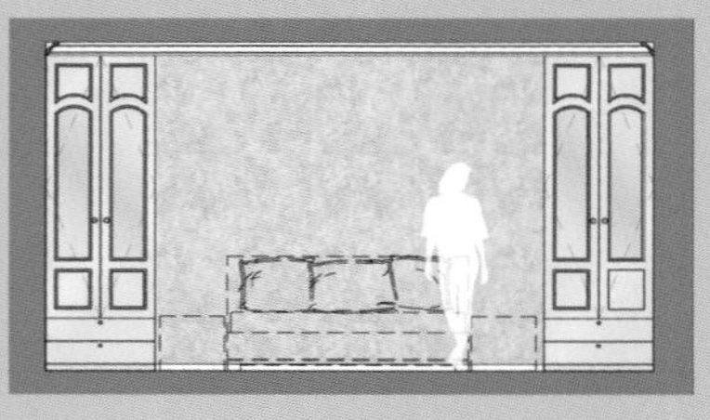

立面图

立面图

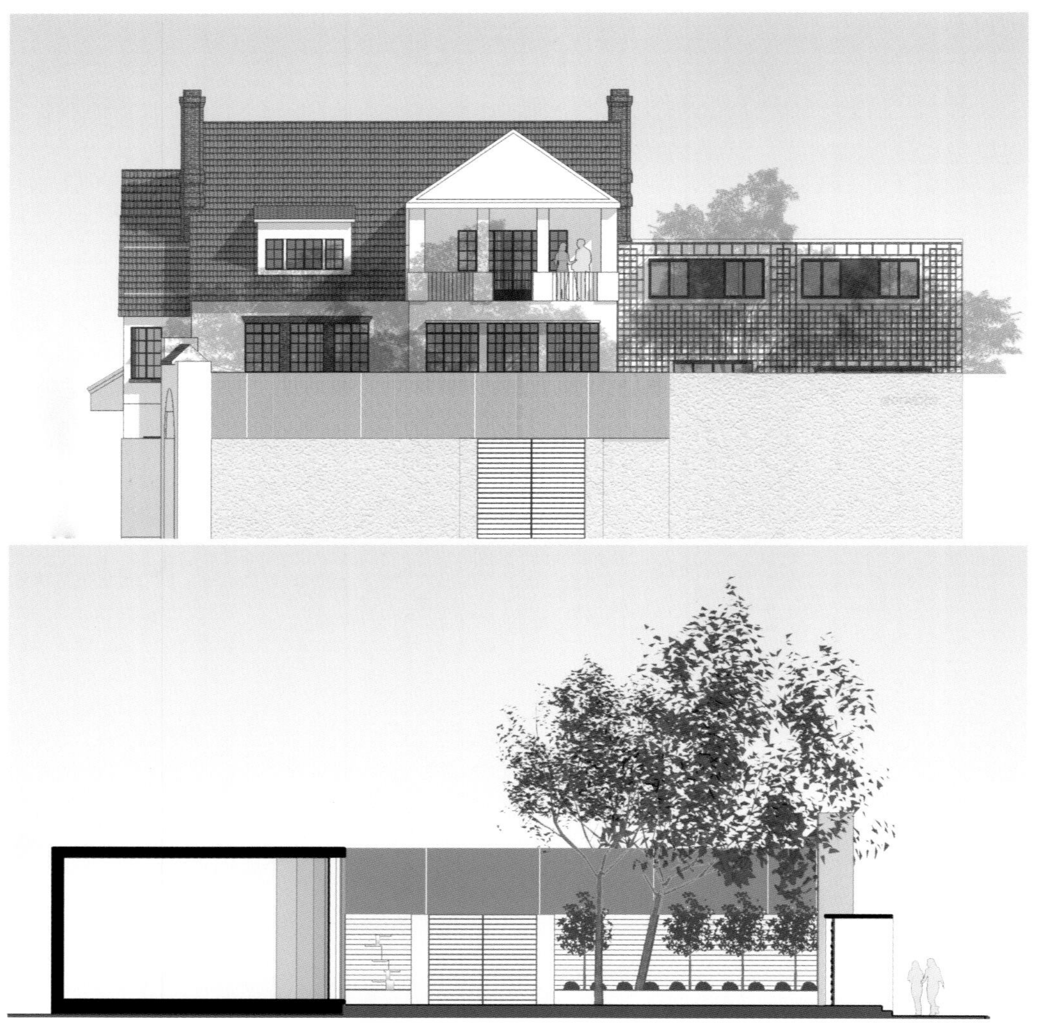

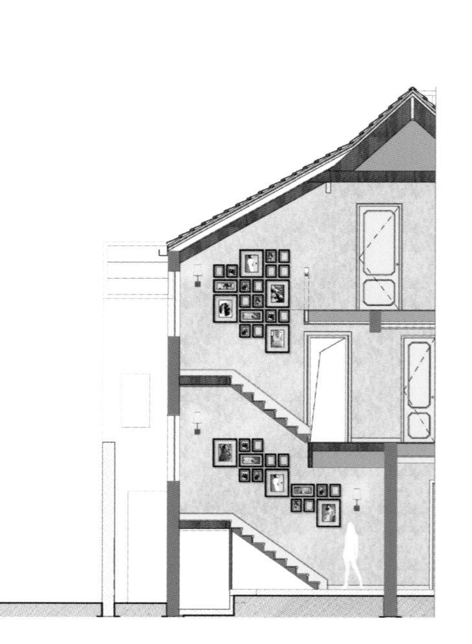

THE OFFICE LADY · THE EXTRAORDINARY FASHION

职场女性 · 非凡时尚

品牌文化 Brand Culture

MOS EDITION

Mos Edition Shanghai showroom is a multi-brand collection store with many individual brands gathered together. The designer is concentrated on creating a store that coordinates all brands in a space. Therefore, on the basis of heterogeneity, SOHO Street where shops of strong individuality gather were chosen as its design concept.

上海Mos Edition陈列室是一个多品牌集合店，许多个性品牌聚集在一起，设计师专注于创建一个所有品牌在一个空间中相互协调的商店。因此，在异质性的基础上，选择了商店密集的SOHO街作为其设计理念。

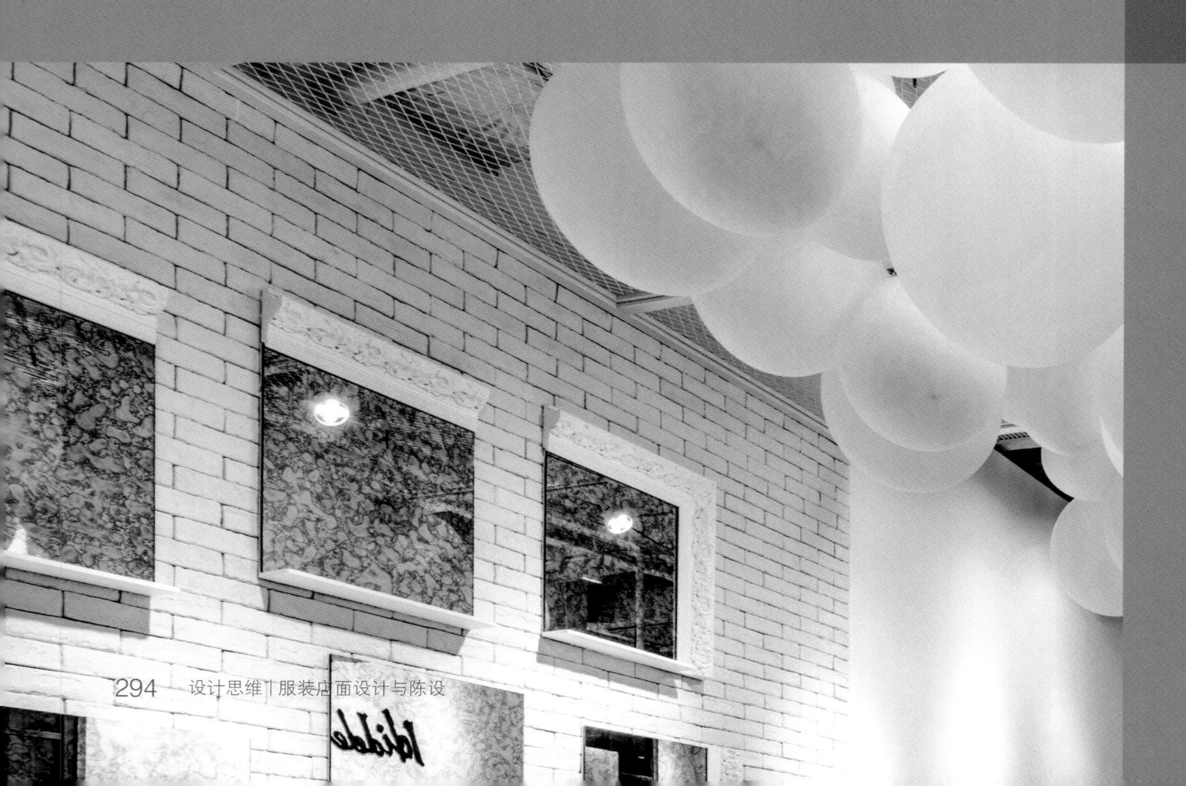

项目信息
PROJECT INFORMATION

扫码查看电子书

项目地点 | 上海
设计公司 | Korea Niiz Design Lab[韩]
项目名称 | MOS Edition女装旗舰店

空间创意

Space Originality

室内陈设
Indoor Furnishings

The designer reveals the individuation of each brand by eliminating the use of colors and decorations in the whole space and by placing "space in space". The showrooms here have the small areas, just like the gable-roofed houses, using the finishing materials with rough texture such as concrete and wood to express the heterogeneous and harmonious feeling. SOHO Street on its interior uses the vivid colors and mosaic tiles. The orange colored glass between the small areas divides human traffic line and leads the visual change.

设计师通过在整个空间中消除色彩和装饰的使用，以及设计“空间中的空间”来揭示每个品牌的个性化。陈列室面积小，就像双坡屋顶的房屋一样，其中综合使用了纹理粗糙的合金材料，如混凝土和木材，同时表达异质感与和谐感。在空间内部的“SOHO街”使用生动的色彩和马赛克瓷砖。橙色玻璃之间的小区域划分了人流的交通动线，并产生视觉变化。

混搭的空间气质

因为客户计划与不同的中国本土时尚品牌合作，所以设计师在不同的氛围里完成了围绕主厅的左右空间。右侧表示SOHO街的异质性和自然性，左侧则通过柜台和吊灯以粗体的形式用光滑的装饰材料表达戏剧性的形象。

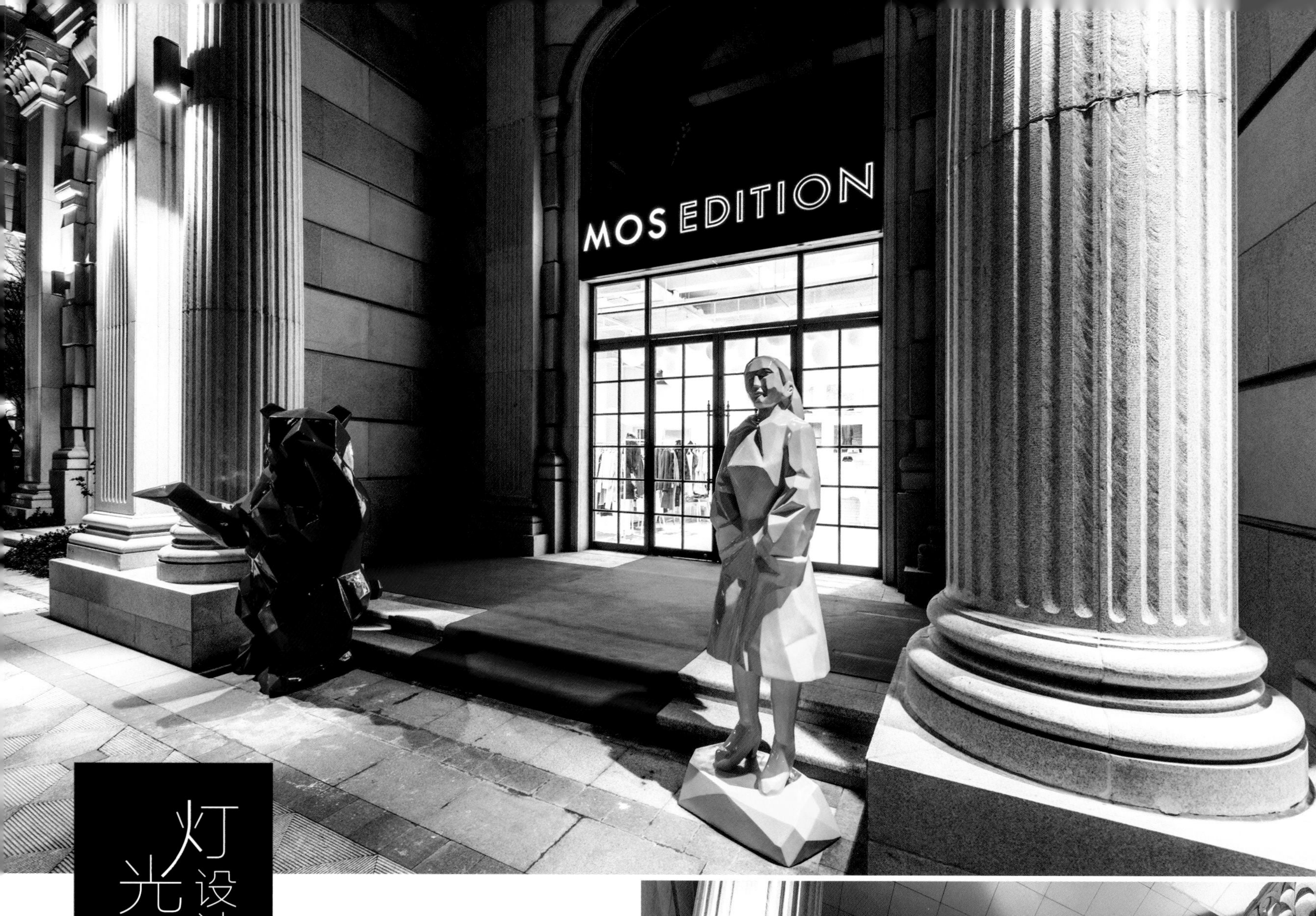

灯光设计
Lighting Design

整体光效

除了陈列室里的聚光效果外，其余空间以白色光源为主，让空间宽敞明亮，顾客可以轻松地欣赏到整个空间的艺术展品及其他设计。

ENJOYING THE LEISURE OF SUZHOU GARDEN IN AKIPELAGO

集岛成群，享苏州园林式悠游

Brand Culture

|AKP|

AKP is originated from the West African IGBO word of AKIPELAGO, meaning archipelago. The brand means that every creator is an island. Based on the concept of its fashion brand SANKUANZ, "AKIPELAGO", founded by the designer Shangguan Zhe, aims to connect the commercial projects of the prominent creators in various fields. The like-minded creators in different fields gather here to create, display and sell their works and share the joy of creation.

AKP来源于西非IGBO族词汇AKIPELAGO，意为“岛群”，品牌取每位创作者都是一座岛屿之意。其由设计师上官喆成立，以自创时装品牌SANKUANZ为理念，旨在连接不同领域杰出创作者的商业项目。各个领域志同道合的创作者在此聚集，共同创作，展示并销售他们的作品，并与大家分享创作的喜悦。

项目信息
PROJECT INFORMATION

扫码查看电子书

摄影师|邵峰
主要材料|玻璃、大理石等
项目面积|1160㎡
项目地点|福建厦门
参与设计|张坚、孙晓楠
主案设计|李泷
设计公司|宽品设计顾问有限公司
项目名称|AKP岛群设计集合店

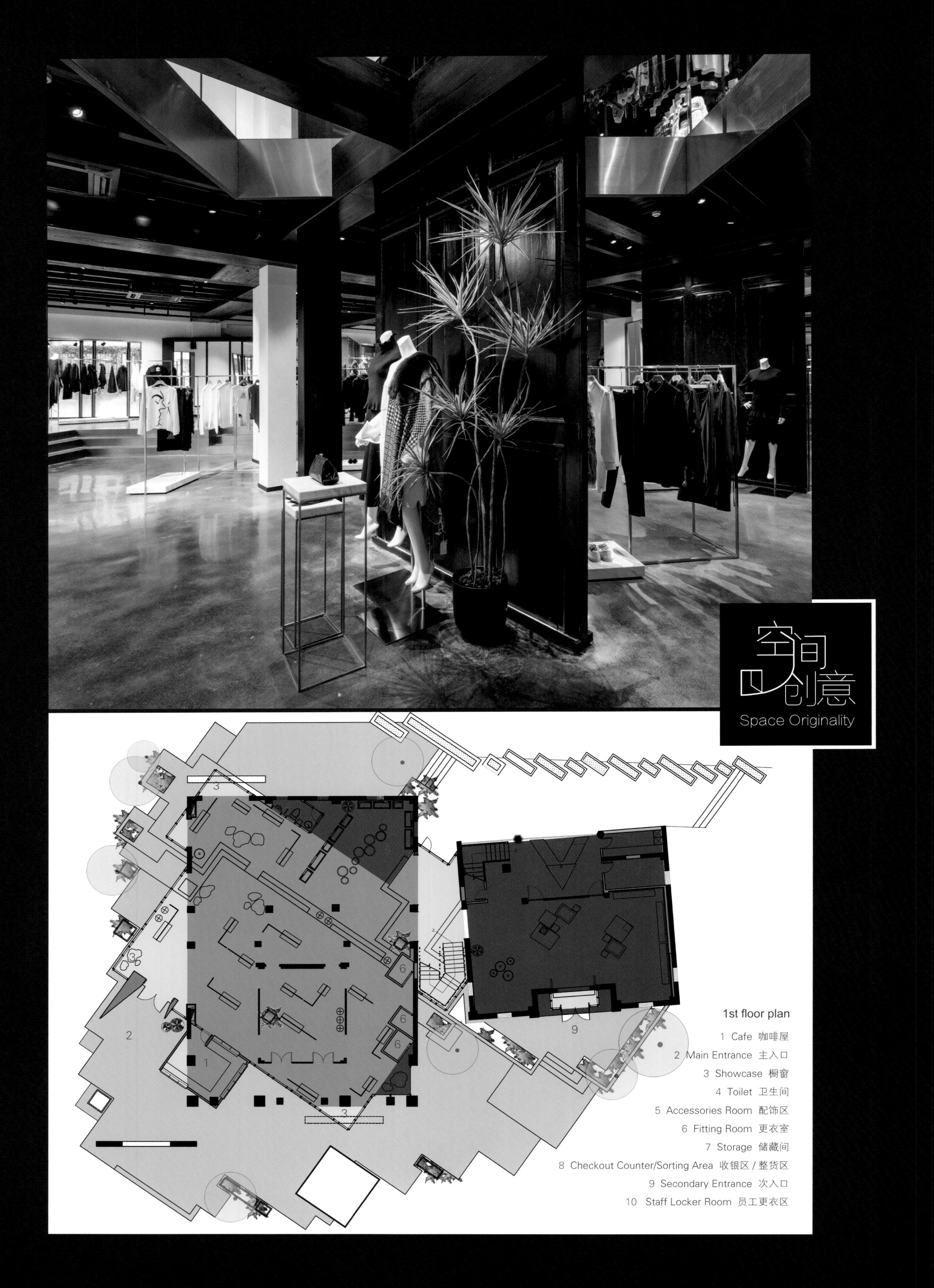
空间创意
Space Originality
3
3
6
6
6
2
1
3
9
1st floor plan
1 Cafe 咖啡屋
2 Main Entrance 主入口
3 Showcase 橱窗
4 Toilet 卫生间
5 Accessories Room 配饰区
6 Fitting Room 更衣室
7 Storage 储藏间
8 Checkout Counter/Sorting Area 收银区 / 整货区
9 Secondary Entrance 次入口
10 Staff Locker Room 员工更衣区

The “AKIPELAGO” project consists of two old building in the period of the Republic of China. The original building is the former residence of the ethnic industrialist, Huang Shijin. Designers use the concept of “parallel universes” to plan the combination of the old and new building, retain and continue the memory and emotion related to the city. The combination of tradition and modernity makes the project sustainable and forward-looking and presents the balance of old and new and the distinct personality of the spatial inclusion and integration.

Designers elevate the public floor outside the building, highlighting the architectural advantages of the “AKIPELAGO” being surrounded and planting tropical plants to connect the islands. The entrance keeps the memorial archway intact and transforms it into an artistic device. With the guidance of the old memorial archway to step into the new entrance of space, the brass portico and the granite archway present a dialogue across time and space. Customers walk on the gray floor, gradually away from the outside noise, purging the impurity of mood, dedicating to space experience. With designers’ careful arrangement of the moving line, customers wander in the opening and closing of the gradual changes in the layout. The moving lines and the horizons are reflected the environmental characteristics continuously. Looking up at the courtyard, it forms a responding with the porch, bringing a profound experience of the combination of old and new and creating a balance.

“岛群”项目由两栋民国时期旧建筑组成，建筑物前身是民族实业家黄世金的故居。设计师以“平行宇宙”的概念来规划新旧建筑的结合，留存并延续与城市相关的记忆与情感。传统与现代的结合使项目具备持续性与前瞻性，并呈现新、旧的平衡以及空间包容与融合的鲜明个性。

设计师将建筑外区域的公共地面抬高外扩，突显“岛群”被簇拥烘托的建筑优势，并种植热带植物串连岛屿印象。入口将牌坊完好保存并转化成艺术装置，通过旧牌坊的指引缓步踏进空间的全新入口，黄铜门廊与岗石牌坊呈现跨越时空的对话。宾客步行于石灰色地坪，逐渐远离外界吵杂声响，涤除心绪杂质，专注地投入空间体验。凭借着设计师精心安排的动线，宾客游走于开合渐变的布局中。动线与视野不断复合环境特色，天井与门廊互为回应，给人带来新旧为邻的深刻体会，促成空间与人、新旧环境的平衡。

[1] 连接老建筑，注入人文底蕴

设计师以“群集的岛屿”为意向进行构思，连通了原本独立的两栋传统建筑，根据不同的功能分区，强化完整的展示概念。使用大面积的水洗石、镜面和黄铜，营造出理性的现代美感，并鲜明对比老宅的粗砺肌理，给现代化的商业空间引入一份文化气息。

[2] 用动线创造游赏趣味

利用中式的装饰窗、柱体和墙体，空间规划回圈式动线，为宾客建构不断亲近园景的行进仪式，犹如在苏州园林中悠游。在梯阶升展台上漫步，人、景相映成趣。

[1] **映照绿植，点缀生机**

空间中的盆栽绿植既作为陈设的一部分，也能在上方光线的作用下，与打扮素雅的高挑模特相伴，共同给空间更多的活力。

灯光设计
Lighting Design

[2] **联动的光线，促成场内与场外的互动**

空间采用玻璃围合，借由玻璃的透光特质，驱使内外场域发生关系。天井区，二楼的阳光洒透至建筑内部，随着时序与光线的变更，牵动时装展示区的光影表情。

[3] **斑驳之中，使人不容忽视**

空间以暖白光为主，投在木质板上的光束再延伸至时装，传达出受重视的暗示。植物和时装的影子在地面交织，影影绰绰的视觉会使人更乐意沉醉于此。

[4] 场外光增强商店存在感

商店外面部分场地较为开阔，设计师给外立面留有充足的光线，并依靠传统建筑结构，营造出各样的光形，使得商店在逛街的人潮以及鳞次栉比的店面中突出出来，让人过目不忘。

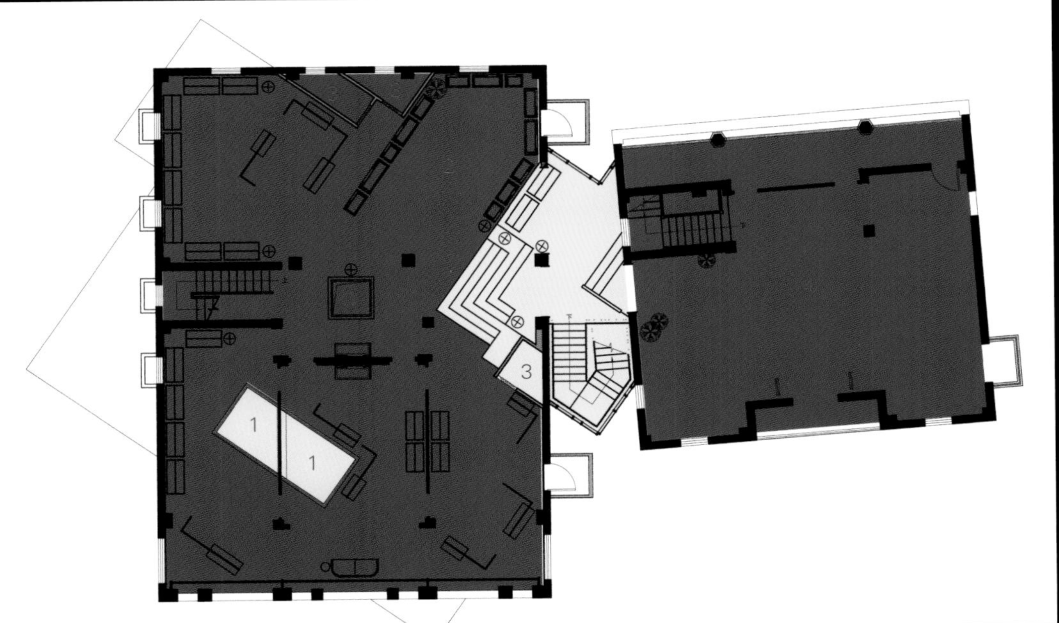

2nd floor plan

1 Patio 上空
2 Accessories Area 配饰区
3 Fitting Room 更衣室
4 Utility Room 备用空间

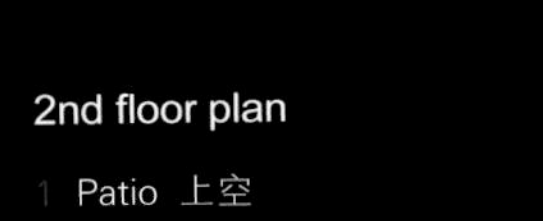

A THREE-DIMENSIONAL MAGAZINE, A LIVING STORE

立体的杂志，活着的店铺

|MAGMODE|

Initiated by the editor-in-chief, the fashion director, the writer and the artist of the international top fashion magazine, Magmode collaborates each season with the fashion publishers, artists and designers from Europe and Japan in order to find and select the most promising designers of the season. And considering the Asian figures, each collection is tailored to fit the Asian figures.

Therefore, Magmode is not only a fashion collection store but also a cultural plan to explore the possibilities of life and will strive to become the birthplace and promoter of the fashion culture through workshops, video, publishing and other media.

SEAN BY SEAN

项目信息

PROJECT INFORMATION

摄影师｜陈兵

项目面积｜600㎡

项目地点｜浙江杭州

设计师｜刘恺

项目合作伙伴｜喆瑞道具

设计公司｜RIGI睿集设计

项目名称｜Magmode名堂杭州嘉里中心店

扫码查看电子书

magmode 名堂
SEAN BY SEAN
choose something else
magmode
本季入选设计师
ARC ATELIER
DISCOVERED
FENG CHEN WANG
F.S.Z
HIUMAN
KENAXLEUNG
MATTITUDE
MEAGRATIA
OLD JOE
SEAN BY SEAN
SCYE
SEAN BY SEAN
01.
atelier
的经典 锐利的精致

Arc Atelier
MATTITUDE
SEAN BYSEAN
SENA

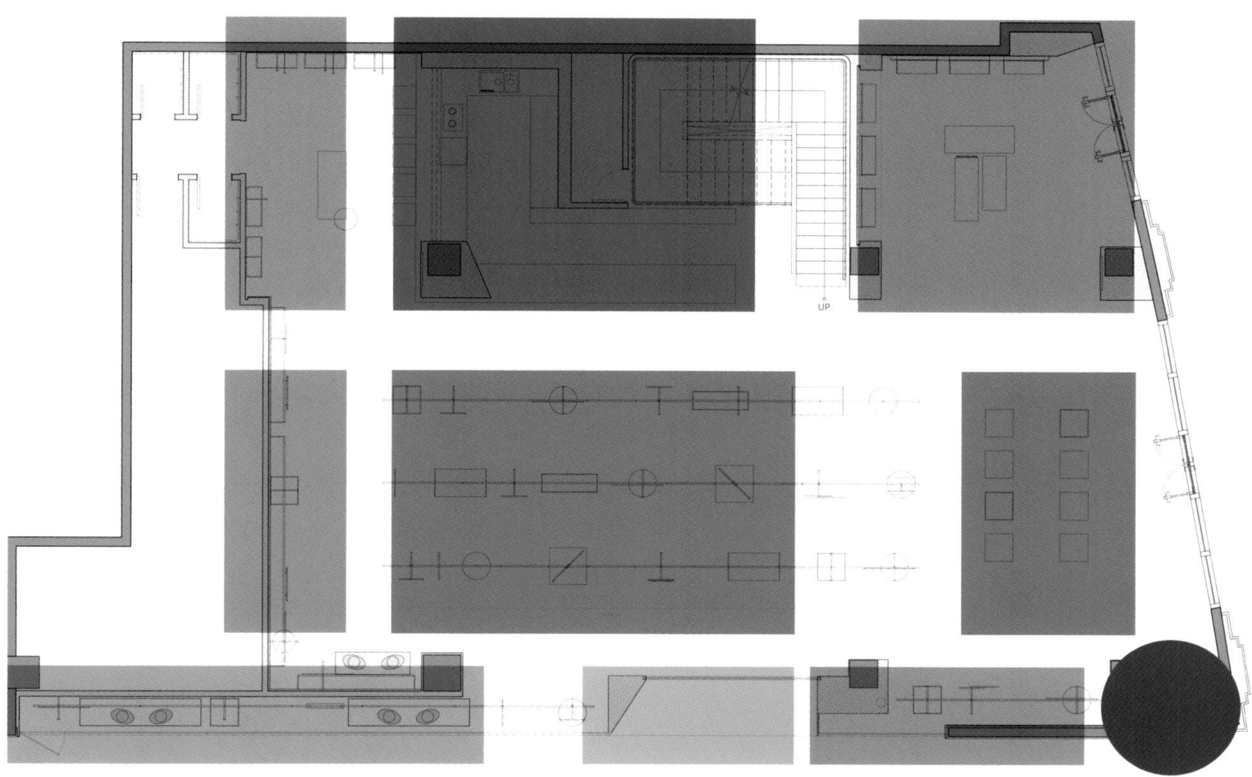

一层平面图

Magmode名堂由国际顶级时尚杂志主编、时装总监、作家、艺术家等共同发起，每季联合来自欧洲、日本等国家的时尚媒体人、艺术家、设计师，在全球范围内挖掘、筛选，最终呈现出这一季最值得期待的设计师，并结合东方人的体型，做出更符合亚洲人的版型调整。

Magmode名堂因此不仅仅是一家时装集合店，更是一个探讨生活可能性的文化计划，并将通过工作坊、影像、出版等介质，戮力成为时尚文化的策源地和推动者。

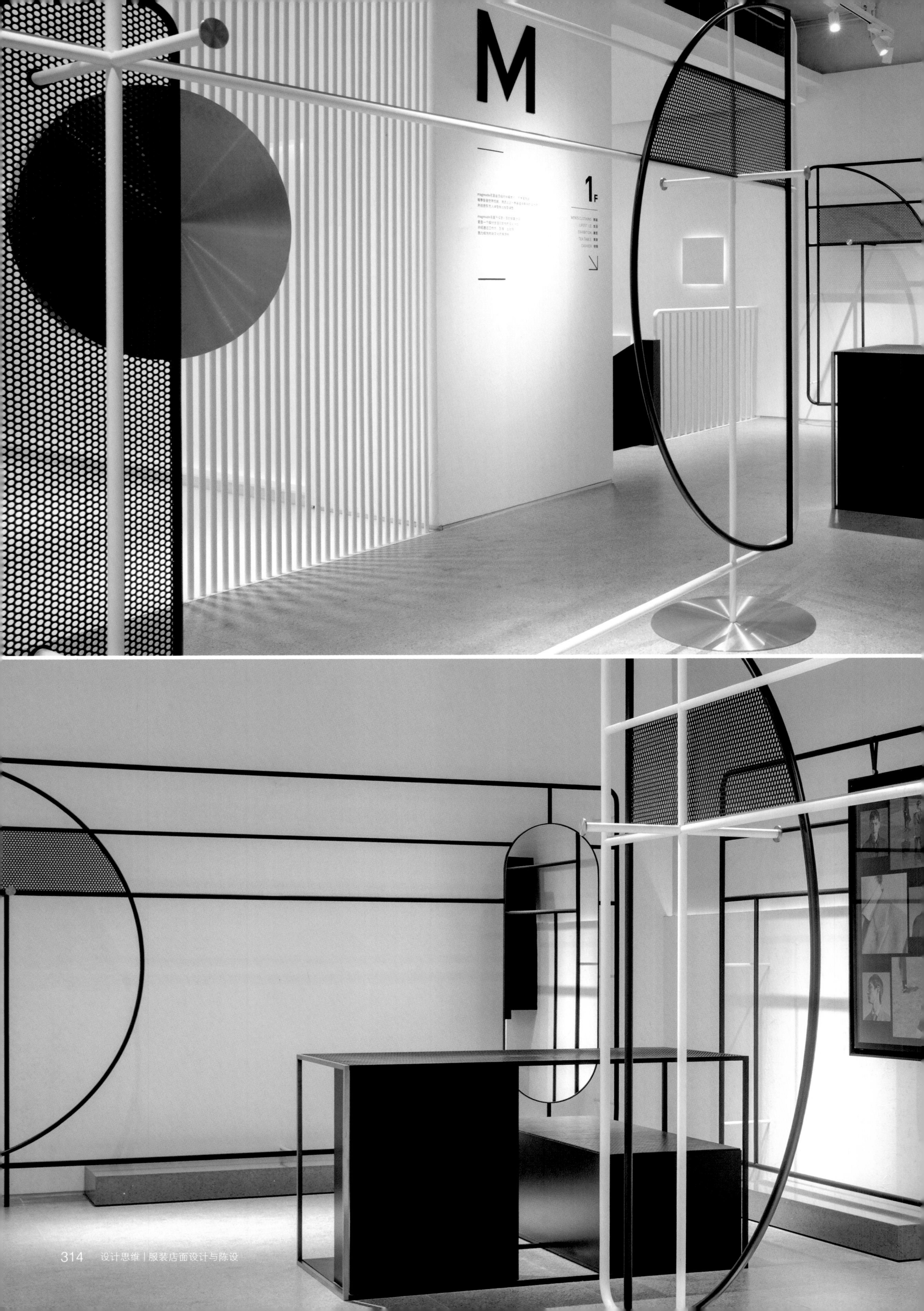
M
1F

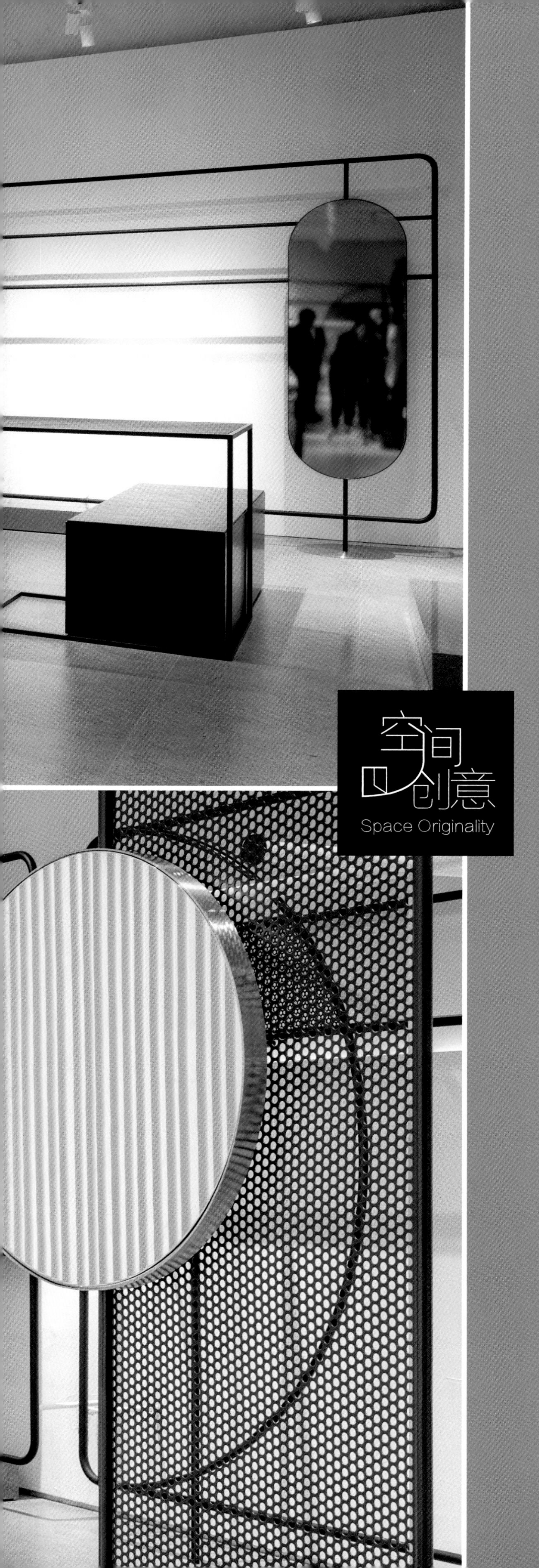

空间创意 Space Originality

There are many ways to express a brand. It could be a monotonous expression or a diversified presentation. It is similar to the magazine in this regard. A magazine has a unified tonality and values, and it starts to contact with readers through different contents, while a brand connects to clients through different products. Its logicality, upgradeability and continuity have something in common with the magazine.

Magmode is a brand that combines works of many designers. Thus a unified concept is needed to express the logicality of the whole brand. In the design of Magmode, the design team hopes to build a new concept in the terminal: a three-dimensional magazine, a readable store.

品牌有多种表达方式，或单一调性地表达，或多元化地呈现，这点和杂志相仿，杂志有统一的调性与价值观，通过不同的内容与读者建立联系，而品牌通过不同的产品与顾客建立联系，Magmode中的逻辑性、更新性、连续性均与杂志有共同点。

Magmode是一个多设计师的集合品牌，需要统一的概念来表达整个品牌的逻辑，设计团队在Magmode的设计中，希望在终端建立一个新的概念——立体的杂志，可以阅读的店铺。

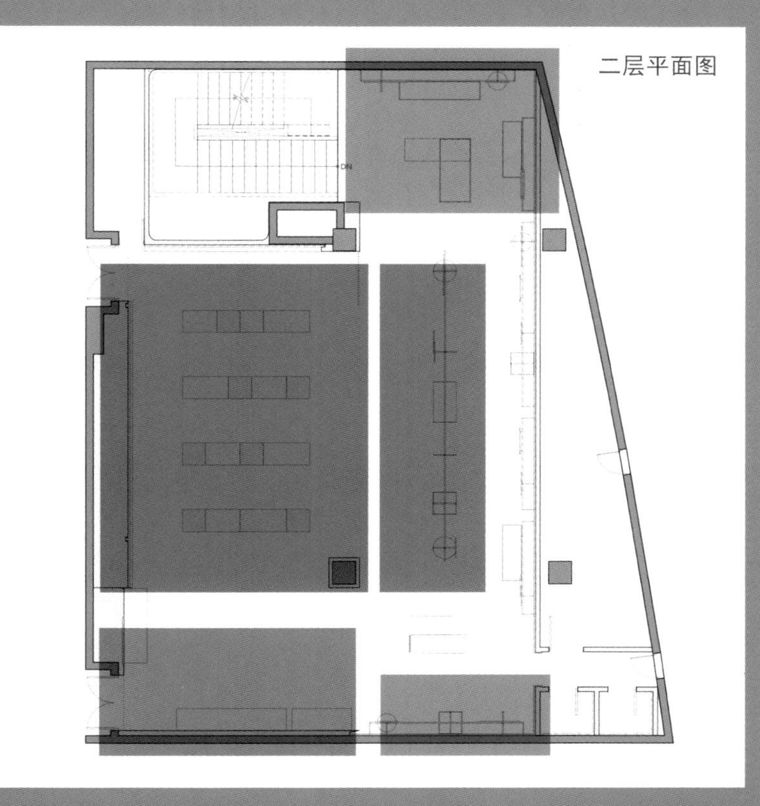

室内陈设
Indoor Furnishings

[1] 精准表达品牌理念

空间是人能在其中获得体验的容器，精准地表达品牌与空间的调性与理念，这是最重要的。空间应该与人发生多元化的交流，即时的新内容会给空间的形态与体验提供更多的可能。

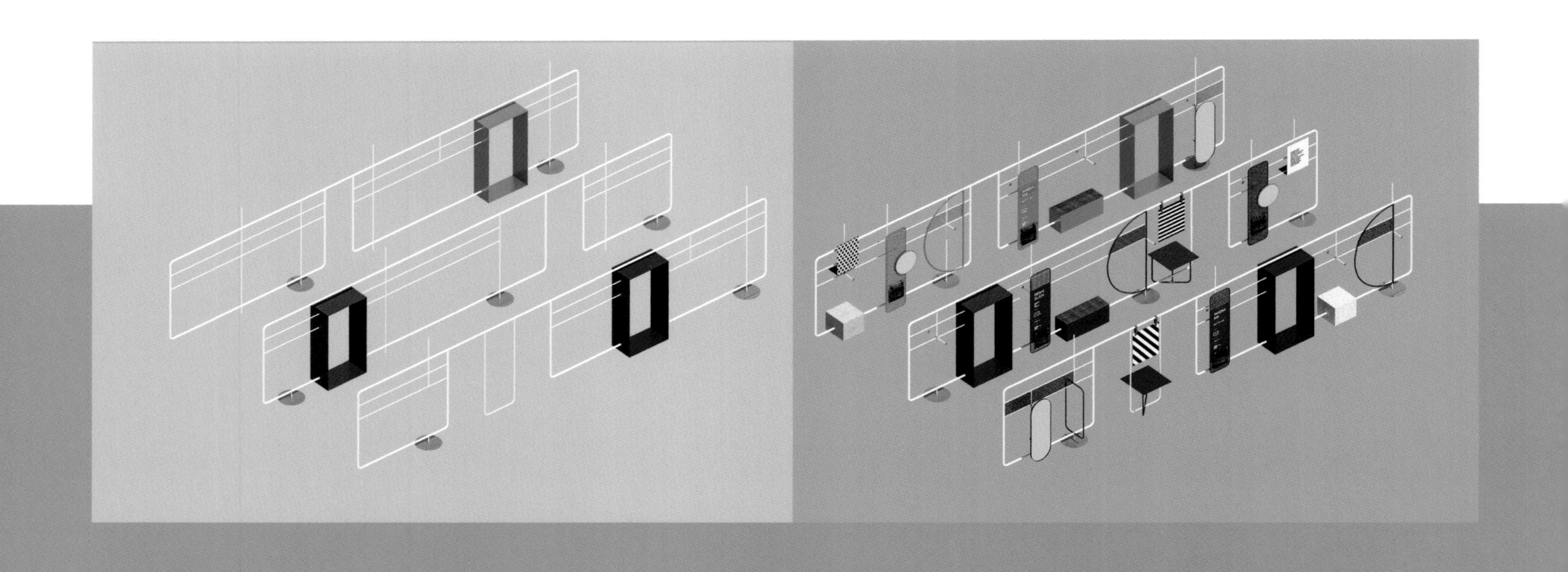

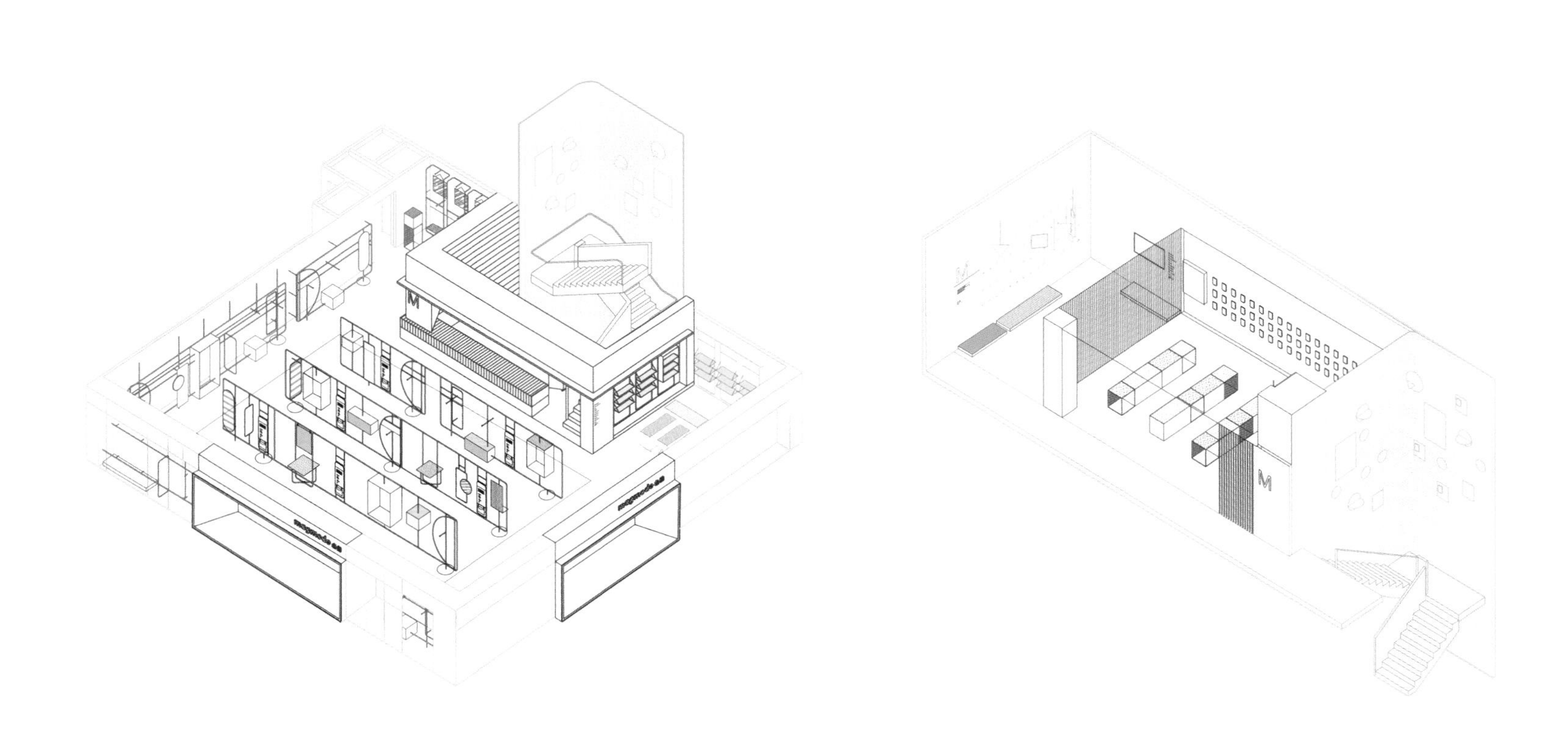

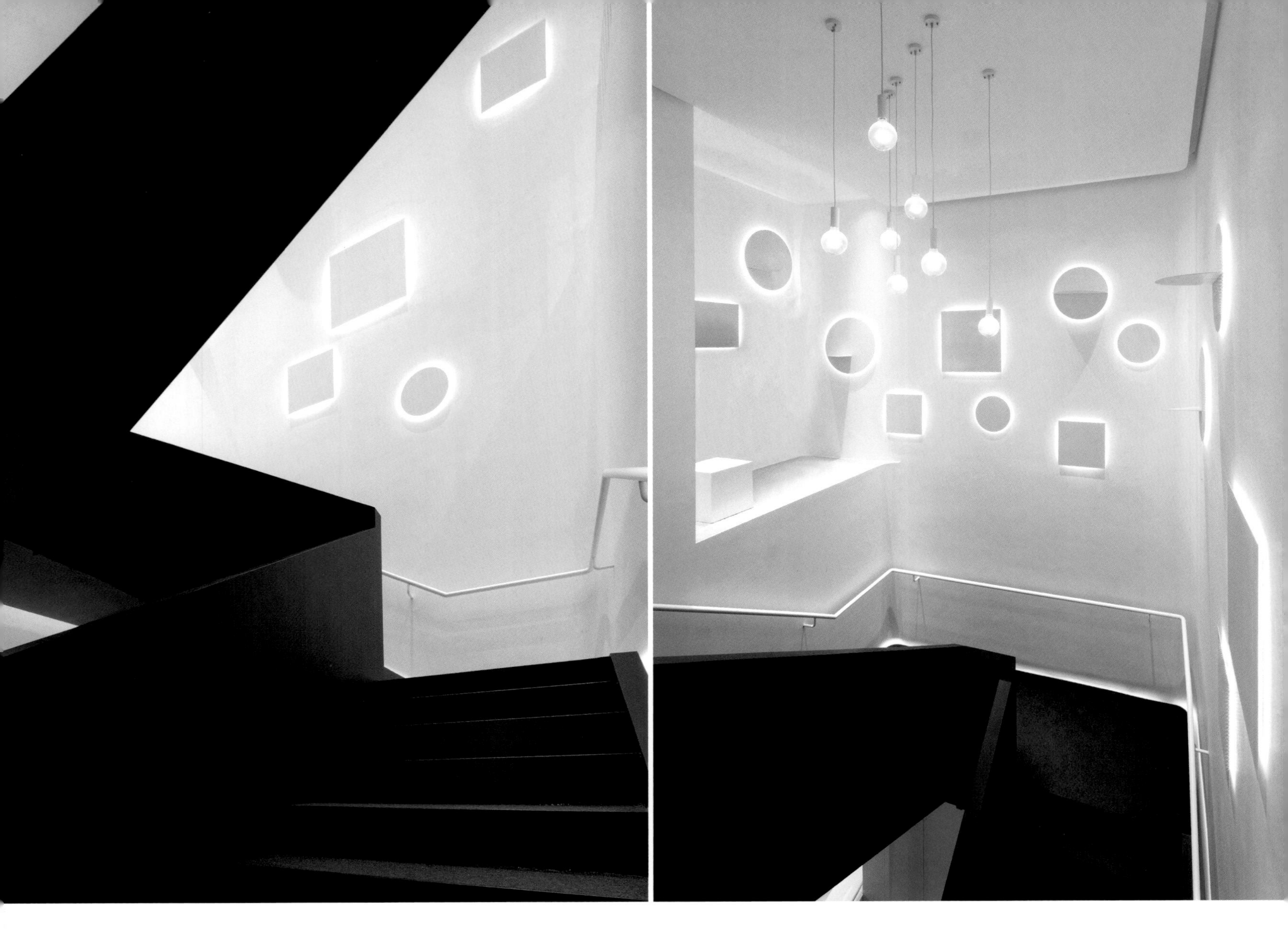

[1] 灯光装饰性

楼梯间的几何图形是黑白空间灵动的装饰，设计师在每一个几何的背后隐藏了光源，让人想象每一个图案都是一个“异次元空间”的出入口。在这里会有非凡的空间体验，而球形的吊灯则增加了纯净空间的轻盈感。

灯光设计

Lighting Design

[2] 细腻的设计

设计师为每一本杂志的展示位都做了一盏照亮它们的灯，让杂志看起来更加立体，成功点题——“立体的杂志”。点状的灯饰、线性的灯光设计，块面的展示区，给人以点、线、面的视觉平衡。

图书在版编目（C I P）数据

设计思维 ： 服装店面设计与陈设 / 深圳视界文化传
播有限公司编． -- 北京 ： 中国林业出版社， 2018.7
ISBN 978-7-5038-9636-1

Ⅰ． ①设… Ⅱ． ①深… Ⅲ． ①服装－商店－室内装饰
设计 Ⅳ． ① TU247.2

中国版本图书馆 CIP 数据核字 (2018) 第 152698 号

出版：中国林业出版社
（100009 北京西城区德内大街刘海胡同 7 号）
http://lycb.forestry.gov.cn/
电话：（010）8314 3518
发行：中国林业出版社
印刷：深圳市国际彩印有限公司
版次：2018 年 8 月第 1 版
印次：2018 年 8 月第 1 次
开本：235mm×335mm，1/16
印张：20
字数：300 千字
定价：428.00 元 (USD 86.00)